Construction Measurements

Wiley Series of Practical Construction Guides

M. D. MORRIS, P.E., EDITOR

Jacob Feld
CONSTRUCTION FAILURE

William G. Rapp
CONSTRUCTION OF STRUCTURAL STEEL BUILDING
FRAMES

John Philip Cook
CONSTRUCTION SEALANTS AND ADHESIVES

Ben C. Gerwick, Jr.
CONSTRUCTION OF PRESTRESSED CONCRETE
STRUCTURES

S. Peter Volpe
CONSTRUCTION MANAGEMENT PRACTICE

Robert Crimmins, Reuben Samuels, and Bernard Monahan
CONSTRUCTION ROCK WORK GUIDE

B. Austin Barry
CONSTRUCTION MEASUREMENTS

D. A. Day
CONSTRUCTION EQUIPMENT GUIDE

Construction Measurements

B. Austin Barry, F.S.C.

Professor of Civil Engineering
Manhattan College

Past National President
American Congress on Surveying and Mapping

A Wiley-Interscience Publication

JOHN WILEY & SONS, New York · **London** · Sydney · **Toronto**

Copyright © 1973, by John Wiley & Sons, Inc.

Library of Congress Cataloging in Publication Data:

Barry, Benjamin Austin.
    Construction measurements.

    (Wiley series of practical construction guides.)
    "A Wiley-Interscience publication."

    1. Surveying. 2. Mensuration. 3. Civil engineering. I. Title.

TA549.B37   624   72-13073
ISBN 0-471-05428-3

Printed in the United States of America

10-9 8 7 6 5 4 3 2 1

Series Preface

The construction industry in the United States and other advanced nations continues to grow at a phenomenal rate. In the United States alone construction in the near future will exceed ninety billion dollars a year. With the population explosion and continued demand for new building of all kinds, the need will be for more professional practitioners.

In the past, before science and technology seriously affected the concepts, approaches, methods, and financing of structures, most practitioners developed their know-how by direct experience in the field. Now that the construction industry has become more complex there is a clear need for a more professional approach to new tools for learning and practice.

This series is intended to provide the construction practitioner with up-to-date guides which cover theory, design, and practice to help him approach his problems with more confidence. These books should be useful to all people working in construction: engineers, architects, specification experts, materials and equipment manufacturers, project superintendents, and all who contribute to the construction or engineering firm's success.

Although these books will offer a fuller explanation of the practical problems which face the construction industry, they will also serve the professional educator and student.

M.D. Morris, P.E.

v

Preface

Every building, road, project, or structure must begin by measurements to map the site so that plans can be prepared. Once the plans are completed, the work must be laid out in the field. One must "set line and grade."

The construction surveyor finds that his function is varied. At one time he will be seeking information to prepare topographic plans for a road, building, or other project. Sometimes he will be doing a survey to establish ground control for aerial mapping or for laying out a project—points to be used to make the transition from plan to reality. At other times, he will be simply maintaining or extending the control survey points on the ground during construction. Sometimes he will be making before-and-after measurements for quantity payments. Finally, when the project is complete, he may be verifying that it is correctly on line and set correctly to grade.

This book provides a means for the field engineer to set out construction staking—the translation of design drawings from paper to the ground. His function is to place on site the controlling stakes, pegs, and information for men and machines to perform construction without a waste of time or effort.

The construction layout engineer-surveyor (who has a basic knowledge of survey instruments and their use, and of the usual surveying operations and procedures) is given the practical knowledge of how to proceed in setting out construction work. Where necessary, theory is given in detail to instill confidence in the methods. New methods and equipment are also described; the use of coordinates is emphasized; current practice is covered; and the descriptions are comprehensible.

It is to be expected that some material is subject to a wider application than that used in any given case. No attempt is made to cover every possible construction situation. I hope that this book will be helpful to the small constructor, the job superintendent, the grading contractor, the construction inspector, and the engineer-surveyor. It will not solve all of their problems, but it is a guide to the solution of some of them, and shows how to do some things. It helps the transition from plans to layout in order to translate the intent of the engineer in charge to the surveyor and contractor who is building the project. Illustrations and explanations cover the common tasks and show how things can be done but, when necessary, further clarification and assistance should be sought. One should always know when to go for help.

B. Austin Barry, F.S.C.

New York, New York
October 1972

Contents

Construction Measurements

Ni 2 employed in the construction of a 220 mW power plant on the Waitaki River,
New Zealand. *Courtesy Carl Zeiss.*

1

Elements of Construction Measurement

1.1   Introduction to Construction Measurement

Construction is the outcome of the civil engineer's design and planning: his project is finally completed. It may be a building, a railroad, a tunnel, a bridge, a canal, a highway, a dam, a complex of many of these, an industrial park, or even an entire community. Intended to serve a particular purpose in a specific place, the project must be laid out in the desired spot, aligned correctly with adjacent structures, placed on the right parcel of ground, and built to the correct size and shape. To lay it out properly, one must take measurements. These pages are devoted to the measurements required to plan, design, lay out, and construct engineered works.

In construction measurement, horizontal distances are measured by tapes, rods, rules, or even roughly graduated sticks. Elevation differences are usually determined by use of a spirit (bubble) level in conjunction with a level rod. Angles are measured mostly by a transit or theodolite, although less accurate instruments can sometimes give adequate results quickly. A sense of the appropriate method and tools is needed; this sense becomes second nature as one acquires experience. Certain new pieces of equipment are becoming available and may very well supersede older, more familiar types. While some of the new equipment is simple, most are expensive and complex.

Care must be taken of the equipment and the instruments used, not simply because they are expensive but also because they must be in good working order if one is to expect reliable results. A good, ingenious technician, however, can often devise simple time-saving methods to accomplish a task without resorting to elaborate or costly equipment. An understanding of the purpose of the work will make the job interesting and guide one in selecting the equipment to do the job.

The use of lineal, elevation, and angular measurements will suffice to locate satisfactorily virtually any point or line or surface for construction. Measurements extending for a great distance may demand that curvature of the earth (the geoid) be considered. Surveys of a limited extent, however, can generally be conducted as though the small surface of the earth under consideration were a flat plane, not a curved surface, thus making the work much simpler.

1.2   Units of Measurement

For *distance*, we commonly use the English system (miles, yards, feet, inches, etc.) although we expect increasingly to encounter the metric system (meters, centimeters, etc.). One must be at home with both systems and be able to relate the one to the other. While the United States, Canada, Great Britain, and English-settled areas are entrenched in feet and inches, the switch to meters will soon be made.

The commonly used units are given below.

English	Metric	
12 in. = 1 ft	1 mm (millimeter) =	0.001 m
3 ft = 1 yd	1 cm (centimeter) =	0.01 m
5280 ft = 1 mi (statute)	1 dm (decimeter)[a] =	0.1 m
1760 yd = 1 mi (statute)	1 m (meter) =	1.0 m
	1 dam (dekameter)[a] =	10.0 m
	1 hm (hectometer)[a] =	100.0 m
	1 km (kilometer)	= 1000.0 m

[a] Not very common.

The relationship between metric and English length units is based on:

$$1 \text{ in.} = 2.54 \text{ cm exactly (since 1959)}$$

$$1 \text{ m} = 3.280839895 \text{ (since 1959)}$$

The fuller explanation of these relationships is in Appendix A.

1.3 Angle Measurement

For *angular* measure, the American system uses degrees, minutes, and seconds, handed down from the "duo-decimal" English system (based on units of 12). On the European continent and in other areas, the *grad*, an extension of the metric system (based on 100), is used. Each system can be understood from this tabulation.

Degree System		Grad System	
Full circle	= 360° (degrees)	Full circle	= 400^g (grads)
One quadrant	= 90° (degrees)	One quadrant	= 100^g (grads)
One degree	= 60′ (minutes)		
One minute	= 60″ (seconds)		

The ratio between degrees and grads is 400:360 or 10:9. For example, a value of 27°15′45″ would become 30.29167^g or 30.2917^g in the metric system, although one need seldom convert. It is, of course, easier to work in the decimal system with no need to handle the sexagesimal conversion of seconds to minutes, minutes to degrees, and so on. But until that time, we are forced to use degrees, minutes, and seconds.

In the degree system, a direction or an angle can be given as 99°, 138°10′, or 87°14′12″, depending on the precision known or desired. Sometimes the value is needed in degrees and decimals of a degree, such as 99°, 138.17°, or 87.2417° in the three previous instances. In making the conversion, the same number of digits in the result will give just about the same accuracy as the original. In converting to the grad value of a degree value, therefore, the same rule can be used.

In the grad system, a decimal system, there is really no such thing as a "minute" or a "second," although some people attempt to clarify things (erroneously) by introducing the terms. For instance, in the grad value cited previously (30.2917^g), they would try to write it as 30^{g}29^{c}17^{cc} and imitate the degree system. This is more confusing than helpful, and one ought to simply learn the grad system as a decimal system and forget minutes and seconds. It is possible that we shall soon adopt the grad

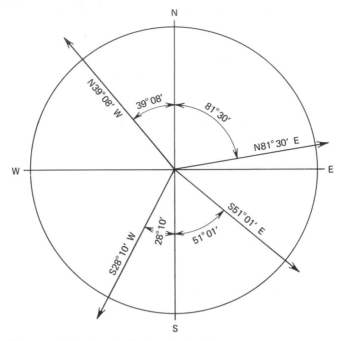

Fig. 1.3.1 Directions by bearings in the degree system.

system, although there are many who believe it will never happen; right now it is strong only in France.

Figure 1.3.1 shows directions of four lines as bearings in the degree system; the bearing angle is always the acute angle measured from the meridian (the N-S line), and the quadrant letters must be attached. Figure 1.3.2 shows directions of the same lines as azimuths in the degree system; the azimuth angle is always the clockwise angle around from the north end of the meridian, and because the "Az" designation is attached or understood, no quadrant letters are used. Figure 1.3.3 shows directions of the same four lines as azimuths in the grad system; the azimuth angle is again the clockwise angle around from the north end of the meridian.

1.4 Azimuths and Bearings

Horizontal angles are measured about a vertical axis, thus being angles in a horizontal plane. Directions of lines in this plane are measured

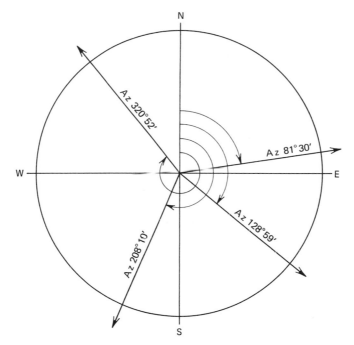

Fig. 1.3.2 Directions by azimuths in the degree system.

clockwise from the true north, an arbitrarily adopted north, or from grid north, and they vary 0 to 360°; they may also be referenced in a particular quadrant, and then will then vary from 0 to 90°. If measured clockwise from the north they are called *azimuths*; if referenced from the meridian in a particular quadrant, they are called *bearings*. One may speak of true bearing or true azimuth, magnetic bearing or magnetic azimuth, local bearing or local azimuth (for the assumed local job), or grid bearing or grid azimuth. One must be able to convert readily from bearing to azimuth and vice versa. For example, the directions of several lines are shown in both azimuth and bearing systems in Fig. 1.4.1.

1.5 Directions and Angles

By subtracting azimuths of lines, one can find the angle between the lines. For instance in Fig. 1.5.1, the angle $COG = Az_{OG} - Az_{OC} = 200° - 75° = 125°$. To find the same angle from the bearings of the

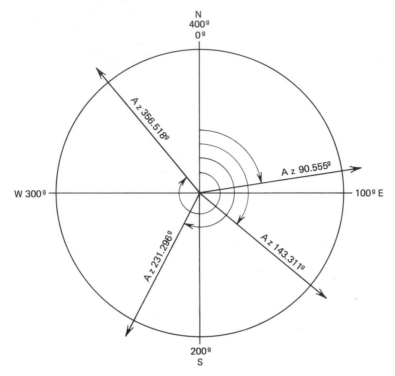

Fig. 1.3.3 Directions by azimuths in the grad system.

lines would call for a sketch (Fig. 1.5.2). It can then be seen which additions and subtractions should be made.

In construction measurement it is needful to know how to handle angular values in degrees, minutes, and seconds. One must know directions, whether bearings or azimuths, and how to do the ordinary adding and subtracting of directions and angles. For example, if *AB* is N71°-15′27″ and the angles are as shown in Fig. 1.5.3, the directions of lines *BC*, *CD*, and *DE* are discovered as indicated on the auxiliary sketches.

1.6   Vertical Angles

The preceding discussion and examples of angles and directions concern horizontal angles, measured about a vertical axis and thus measured in a horizontal plane. Virtually the only other type of angle one need be concerned with is the vertical angle. (See Fig. 1.6.1) Vertical angles are

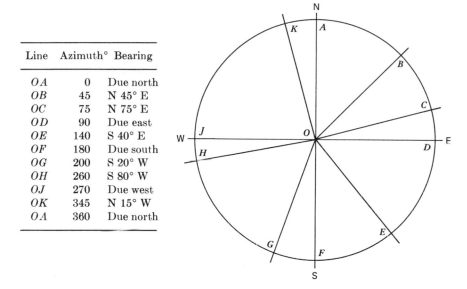

Line	Azimuth°	Bearing
OA	0	Due north
OB	45	N 45° E
OC	75	N 75° E
OD	90	Due east
OE	140	S 40° E
OF	180	Due south
OG	200	S 20° W
OH	260	S 80° W
OJ	270	Due west
OK	345	N 15° W
OA	360	Due north

Fig. 1.4.1 Relationship between directions in bearing and azimuth systems.

measured above or below the horizontal plane as it is determined by a spirit bubble. When the point sighted is above the horizontal, it is an angle of elevation, indicated typically as + 3°10′, or + 11°13′30″, and so on. Some modern instruments, however, to avoid the mistakes that can occur with the + (plus) or − (minus) signs, use 90° as the horizontal, or level sighting.

If an instrument has the vertical circle reading 0° when pointed upward to the zenith and 180° when pointed directly downward, it gives "zenith angles." Then an angle of 93°10′ would be a negative vertical angle of −3°10′ (an angle of depression), and a zenith angle of 86°50′ would be a positive vertical angle of +3°10′ (an angle of elevation).

If an instrument has the vertical circle reading 180° when pointed upward to the zenith and 0° when pointed directly downward, it gives "nadir angles." Then an angle of 93°10′ would be a positive vertical angle of +3°10′ (an angle of elevation), and a nadir angle of 86°50′ would be a negative vertical angle of −3°10′ (an angle of depression).

When taking notes in the field, care must be used to indicate whether the angles are conventional, zenith, or nadir angles so that the office people will not make a mistake. All three types of instrumentation are currently in common use.

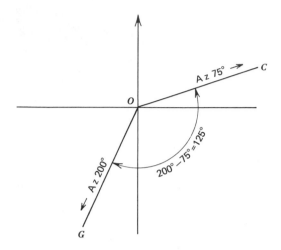

Fig. 1.5.1 Angle from azimuths.

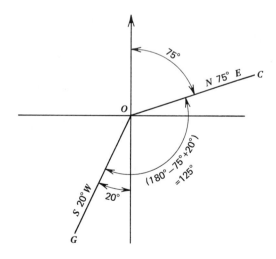

Fig. 1.5.2 Angle from bearings.

8

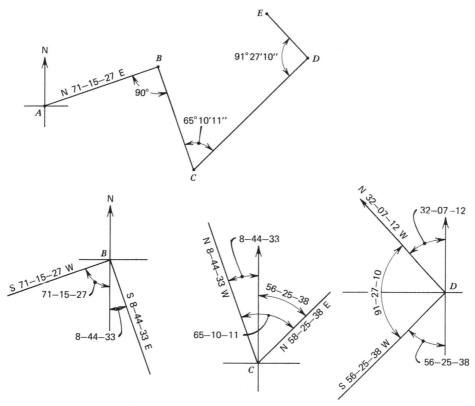

Fig. 1.5.3 Deriving directions from angles.

1.7 Accuracy and Precision

A word about "accuracy" and "precision" is in order, because throughout the text these terms will underlie the work being described. Accuracy is nearness to the true, closeness to the right answer. In tunneling, for example, meeting exactly in the middle of the mountain will prove the work has been accurate. Precision, on the other hand, is refinement of measurement, closeness of one measurement to another. Precision is highly desirable and does bring about accuracy, although of itself it is no guarantee that the two parts of the tunnel will meet. Precision is a function of the equipment's readability and of the operator's skill, and is seen in the fineness of the resulting measurements. Accuracy can only be defined in terms of the trueness of the result, its closeness to the correct value. One can achieve both precision and accuracy through the use of

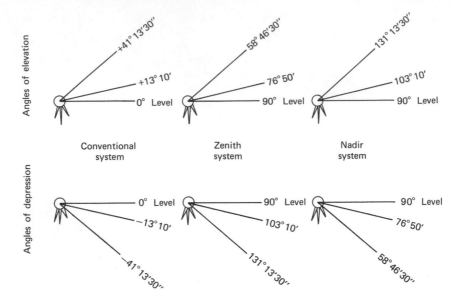

Fig. 1.6.1 Comparison of vertical angle measurement systems.

care, good instruments, and correct techniques, plus expenditure of time and patience. One can estimate accuracy by examination of the precision of the work as it progresses, but finally only by examination of the end result. Independent check measurements will build up confidence in one's accuracy; so, too, will experience.

1.8   Errors and Blunders

A note about errors and blunders is also in order. Compensation for systematic errors is possible, and must always be done. These are errors whose size and sign are knowable and should be known. There are other types of errors, called accidental errors, whose size (usually small) and whose sign cannot be known. These cannot be corrected, but they can be minimized with use of precise equipment, working with systematic procedures, and using care. Blunders are mistakes, pure and simple, and are never excusable. A surveyman should correct systematic errors, for they are compensative; with care, he should minimize accidental errors, and never be guilty of blunders.

Treatises on errors exist, and should be consulted. From such study one can determine the validity of his result from statistical theory and probability. It renders possible such statements as, "The point placed here is correct to ± 0.04 ft." It makes one capable of guaranteeing one's results with a 68, 90, or 99% probability, because one has followed a particular procedure.

A basic notion is that construction measurements must be reliable and free of blunders. Cross-checking as one progresses tends to provide needed assurance, and continual care to verify results is demanded. Because important work hinges on the information acquired or set out in the field, a fairly high monetary value must be associated with the work, along with an appropriate sense of responsibility.

2

Difference of Elevation

2.1   Elevation and the Reference Datum

Man's sense of difference of elevation is instinctive; we know uphill and downhill, we keep tables level and hang pictures straight, we observe

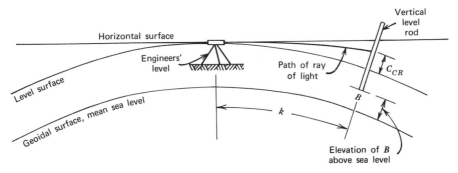

Fig. 2.1.1 Curvature and refraction.

still water surfaces, and so on. We speak of "elevation above sea level" quite knowingly. The practical matter of measuring from the sea level base or *datum* is somewhat difficult, however, especially to any great distance inland.

The geoidal surface of the earth (the geoid) is simply regarded as the level of the mean ocean surface as if it were extended underground all around the earth. The commonly used datum is this geoidal datum or mean sea level surface, everywhere perpendicular to the direction of gravity. It is true, though, that many other datum surfaces are locally employed. Mostly these local datum planes were in use before good sea level measurements were available in the locality. Care must always be exercised to know which reference datum is being employed.

Because of the curvature of the earth and the refraction or bending of light rays, the path of the ray of light departs both from the horizontal plane and from the level surface, as seen in Fig. 2.1.1. In any situation employing fairly long level sighting (with the engineers level or any telescopic instrument), a correction to the rod reading may have to be employed. The formula for the correction for curvature and refraction is:

$$C_{CR} \text{ (in ft)} = 0.572k^2$$

where k is the sighted distance in miles. The ray of light differs from the level line by this amount.

2.2   Difference of Elevation

Measurement in the vertical direction to find difference of elevation is simple enough if the two points are directly one above the other. If not,

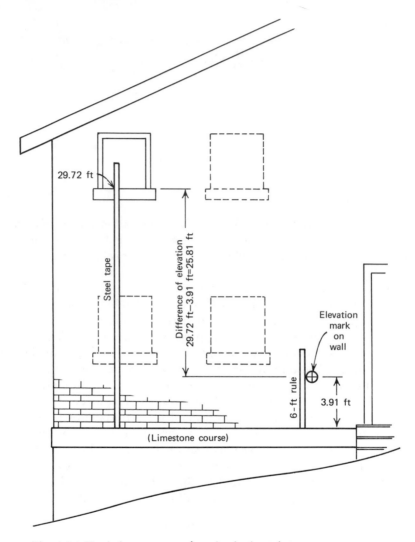

Fig. 2.2.1 Vertical measurement by using horizontal stone course.

then a horizontal plane is used for reference, with the measuring being made up from the plane to the two points and the proper addition and/ or subtraction employed. The reference plane can be a real horizontal surface, such as a floor, a level string, a window sill, or a brick course, as in Fig. 2.2.1, or an imaginary plane. The most common imaginary plane is one fixed by sweeping a level line of sight about a vertical axis as determined by a spirit level bubble.

2.3 Simple Spirit Bubble Levels

There are also two simple and relatively rough leveling devices in common use that rely on a spirit bubble: the string level and the carpenter's level. These are shown in Figs. 2.3.1 and 2.3.2. The idea of "spirit" is simply that to prevent freezing of the liquid, alcohol was formerly used; today a refined petroleum is placed in the level vials to form the bubble.

2.4 Water-Tube Levels

Results obtained with these small leveling devices are frequently satisfactory for many construction needs. Another simple device, dating back

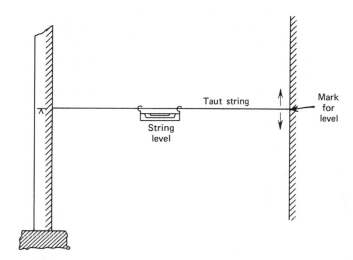

Fig. 2.3.1 Use of string level.

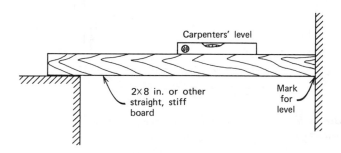

Fig. 2.3.2 Use of carpenter's level.

to antiquity, is the water tube. In its simplest form it can be a garden hose filled with water and equipped with a glass or plastic sighting tube at each end. In Fig. 2.4.1, the elevations of two piers are compared so that the structure can be shimmed to be level when it is set on the piers. There is a more elaborate water-tube assembly that is manufactured for semipermanent mounting on a building. Rather exact readings can be made to detect small settlements in critical construction; Fig. 2.4.2 shows the micrometer gauge at one reading station.

The water level assembly is used in construction work and industrial applications where it is not convenient to use conventional methods of leveling. For example, in a construction excavation, shoring timbers and sheet piling may make it impossible to use surveying instruments. In this case, the water level can be used to determine elevations or changes in elevation.

The most frequent use of this water level is to measure differential settlements in structural foundations in buildings, power plants, or dams, or in precise leveling required for heavy production machinery, or electric and hydraulic generating and turbine equipment. The measuring points can be installed in walls or columns, interconnected by lengths of flexible plastic tubing filled with water (or saline solution). Reading of water level is accomplished by a steel micrometer spindle in a reservoir assembly at each point to be controlled. Measurements to 0.005 in. are readily obtained.

A use of the water level as a control device in slip-forming a tall building is illustrated in Fig. 2.4.3. At each jack location, visible water levels assist the jack operator to control the form's upward movement. The entire slip-form assembly is kept level during the jacking, thus preventing

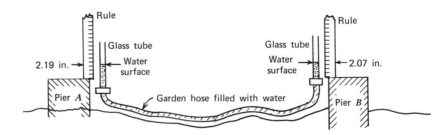

Fig. 2.4.1 Use of water-tube level to compare relative elevations of two piers during building construction.

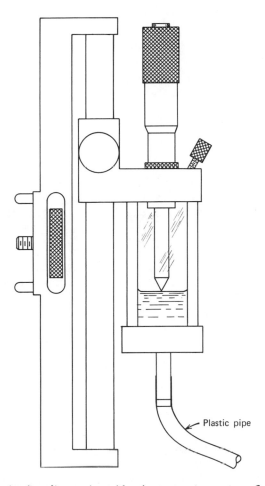

Fig. 2.4.2 Water level reading station with micrometer gauge.   *Courtesy Soiltest, Inc.*

it from deviating to one side or another and pouring the concrete wall out of plumb. In fact, the method employed shows the individual jacks on the four corners being controlled by limit-switch floats riding on the water in each container. If one jack rises too fast it gets shut down for a short time. Of course, to supplement this water-tube leveling method, occasional precise spirit-level instrument readings on each of the four corners are a precautionary check to control elevation accuracy (story height) and thus assure that the structure will be plumb.

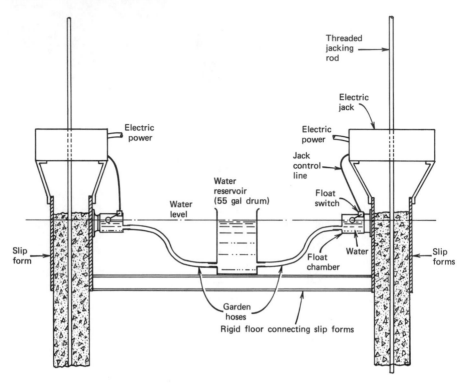

Fig. 2.4.3 Scheme for maintaining slip-form level and plumb while jacking on a concrete silo or building.

2.5   *Water Surface Leveling*

A larger version of the water-tube level is the use of the water surface of a still lake or pond. Transfer of level across such a body of water can save time and effort, as long as tidal action, current, or piling-up of water through wind action does not give erroneous results. It is applicable as well to estuary and swampy delta situations where very slow and slight movements of water may be occurring, although caution must be used. This would call for measurements over an extended time at periodic intervals so that assured correlation can be obtained. However, the effort could achieve an otherwise nearly impossible result with a fair degree of accuracy and with relative ease.

The method in Fig. 2.5.1 shows the rod being read at the water surface by using a leaky or perforated can to smooth out the ripples at the same time that the rod is read by the instrument. Simultaneous measurement on the opposite shore—on several different occasions, probably—would be advisable for reliability.

2.6   The Hand Level

The hand level, shown in Fig. 2.6.1, is a nonmagnifying tube fixing a line of sight, with a spirit bubble attached. The observer sees the target and the bubble simultaneously while holding the instrument to his eye (perhaps braced against the ground by a range pole for added rigidity). It is not a precise instrument, but it can serve adequately for many simple tasks. (See Fig. 2.6.2) Normally with a hand level, one reads a level rod, a carpenter's rule, or a simply marked stick for determining differences of elevation for a quick and rough result. In setting slope stakes on a highway or giving a rapid reading for construction grading, for example, it can serve a real purpose.

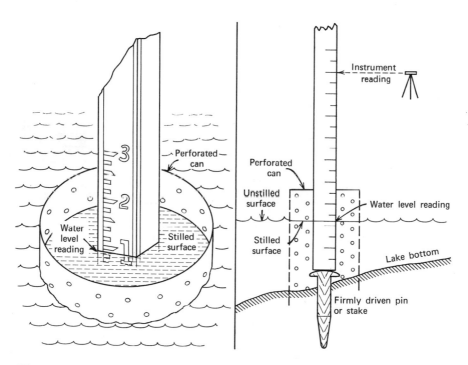

Fig. 2.5.1 Elevation transfer from stilled water surface.

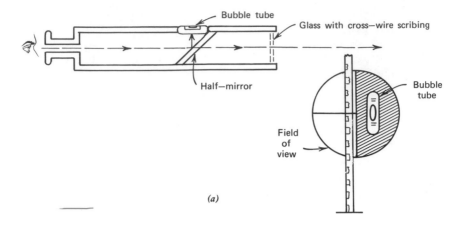

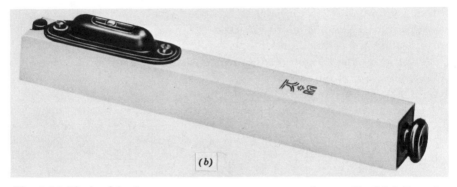

Fig. 2.6.1 The hand level. *Courtesy Keuffel & Esser Co.*

Fig. 2.6.2 Use of hand level to compare elevations.

2.7   The Abney Level

A variation of the hand level is the clinometer or abney level, in Fig. 2.7.1, equipped with a small graduated circle and having its spirit bubble movable with the circle. This device can be used to check an inclined slope, as in Fg. 2.7.2, or a drainage pipe or ditch, and so on, in a simple but sometimes satisfactory manner. It can be used as a sighting device or it may be laid down on a smooth inclined surface. The inclination of

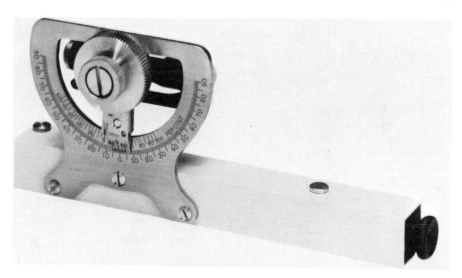

Fig. 2.7.1 Abney level.   *Courtesy Keuffel & Esser Co.*

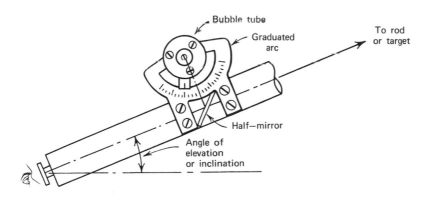

Fig. 2.7.2 Abney level used for vertical angle.

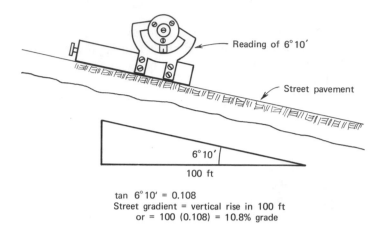

tan 6° 10′ = 0.108
Street gradient = vertical rise in 100 ft
or = 100 (0.108) = 10.8% grade

Fig. 2.7.3 Abney level used to obtain slope of pavement. The reading in degrees may be converted into percent slope by use of tangent tables.

the line of sight or of the surface will be read in degrees when the level bubble is centered. A street gradient could be thus obtained, as shown in Fig. 2.7.3.

2.8 Checking the Simple Spirit Level

A hand level or a clinometer can be checked easily by the technique of reverse reading, as in Fig. 2.8.1. The level is held against a level rod *A* at, say the 4.5-ft mark, and a reading on distant rod *B* is taken, which may be, say 3.8 ft. Then the level is held against rod *B* at 3.8 ft, and the sighting on rod *A* should give 4.5 ft. If not, the level is inclined up or down by some amount, and an adjustment by the cut-and-try method must be made. In some instruments, the cross wire is loosened and moved up or down; in others, the interior prism is shifted forward or back. Successive adjustments and trials are made until the level reads correctly. The string level or the carpenter's level or any simple bubble level is checked by reversing its direction in much the same manner. A taut string is held level as the bubble of the string level indicates; then the string level is turned end for end on the string and the bubble should again center. If it does not, one of the end hooks is raised or lowered (or bent slightly) until the bubble centers in both directions on the string. A carpenter's level is usually not adjustable, but a check on its accuracy

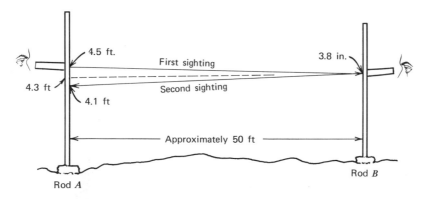

Fig. **2.8.1** Checking the simple spirit level.

can be made. A smooth board is set so the bubble centers; then the level is turned end for end and the bubble should center. If not, another level should be obtained.

Remember, however, that precise results cannot be obtained with the hand level, the string level, or the carpenter's level. They suffice, though, for ordinary work in setting concrete forms, brickwork, house framing, and so forth.

2.9 Barometric Leveling

An infrequently used but quite rapid method of ascertaining elevation is the aneroid barometer, which gives elevations through measurement of atmospheric pressure. Airplanes regularly use such barometric devices, called altimeters. On land, one can carry a fairly accurate altimeter, set it down a moment, and read the elevation to within a few feet. This is called the surveyor's altimeter, shown in Fig. 2.9.1.

It is necessary first to set the altimeter on a point of known elevation and adjust it to read that elevation. It is then carried to any other points where an elevation is wanted, and at each such point an elevation is recorded—as well as the time of the reading. Normally, since atmospheric pressure does change perceptibly during a short period, a second similar altimeter is kept at the beginning station (of known altitude) and reference readings are taken there at regular intervals during the work. The time is noted at each observation of the roving altimeter and in this way, corrections can be made to the elevation readings that are obtained by

Fig. 2.9.1 Surveyors altimeter.   *Courtesy Keuffel & Esser Co.*

the rover. The last observation by the roving barometer should then be made at the initial station as a check. For quick reconnaissance a fair-quality set of elevations can be thus obtained for initial planning for a large area where more cumbersome methods could not be economically justified.

2.10 The Telescopic Level

The most common and most accurate device is, of course, the engineer's level. The engineer's level is a telescopic line of sight to which is affixed a spirit bubble, tripod-mounted, which can sweep a horizontal plane to intercept a rod held vertical. One can read vertical difference of elevation, depending on the instrument, ranging from ordinary (±0.01 ft) to precise (±0.001 mm). Several varieties are made and the correct instrument should be selected for any given task. Basically, the telescope enlarges the field of view for more accurate readings, enables refined sightings, and is a more stable device. Figure 2.10.1 shows a typical engineer's level.

The tilting level of Fig. 2.10.2 is an engineer's level equipped for more rapid setup and reading by means of the tilting axis.

Fig. 2.10.1 Engineer's dumpy level. *Courtesy Teledyne Gurley.*

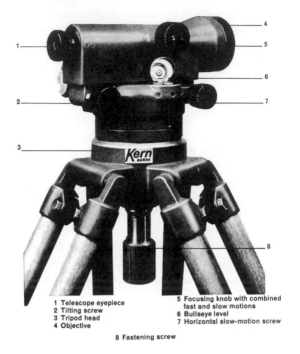

1 Telescope eyepiece
2 Tilting screw
3 Tripod head
4 Objective

5 Focusing knob with combined
   fast and slow motions
6 Bullseye level
7 Horizontal slow-motion screw

8 Fastening screw

Fig. 2.10.2 Engineer's tilting level.   *Courtesy Kern Instruments.*

The constructor's or contractor's level is a simpler form of the engineer's level and serves adequately to set elevations on a small building site or whenever great accuracy is not critical. It suffices for most construction, especially serving as a less expensive supplement to more precise types, on the day-to-day marking and checking of elevations. Figure 2.10.3 shows a combination constructor's level-transit.

The more precise level (engineer's) can be employed for work demanding less precision, but the less precise level (contractor's) cannot be used for very precise work.

2.11   Precision in Leveling

Various degrees of precision can be employed in ascertaining elevation differences, from rather crude string leveling to sophisticated precise engineer's leveling. Leveling in order to install a cyclotron, for example, is quite different from leveling the foundation for a simple frame house;

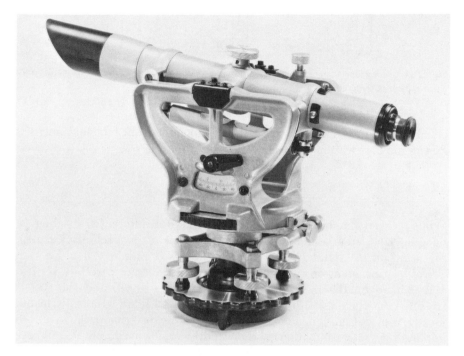

Fig. 2.10.3 Constructor's transit-level instrument.

the former needs very precise equipment, while the latter does not. Judgment must be employed to know how exacting one should be in any given situation. If long sights are employed (over 150 or 200 ft), the variation of a level line from a horizontal line (caused by earth curvature) may have to be considered in the level procedure. This was discussed in Section 2.1.

The ordinary engineer's level (dumpy level) has a rather sensitive bubble that is centered by the leveling screws. If the bubble is centered before and after each sight, one may conclude that it was centered during the sighting. It is common on the tilting level to have a mirror prism for viewing the bubble tube at the instant of sighting the rod. One form is the coincidence reading bubble, which splits the bubble so both ends can be observed almost while the sighting is being made. (See Fig. 2.11.1.) One uses this instrument in the following manner. Focus the telescope on the rod. Turn the tilting screw until the bubble ends are in coincidence and the line of sight is thus horizontal. Take the rod reading, then check the bubble coincidence immediately. If the bubble ends coin-

Fig. 2.11,1 Coincidence or split bubble showing opposite ends of bubble coincide only when bubble is centered.

cided both before and after the observation, they must have coincided during the observation.

2.12 Level Rods

While it is possible to use any graduated scale for differential leveling (a hanging tape, a carpenter's rule, a marked stick, or a machinist's scale), the usual device sighted is an especially devised rod, which is held on a firm bearing point on the ground. Its graduations are carefully scribed and painted on the rod, with good visibility in mind, and the rod is sturdily constructed for durability. In the United States, the rods in use mostly use feet and decimals of feet, not inches; graduations are to hundredths, with estimation to thousandths easily accomplished by eye or by use of a vernier target. (See Fig. 2.12.1.) However, the level rod can equally well be used to do less precise leveling, at which time it is not unusual that the readings be made only to hundredths or to tenths of feet. When using a hand level, for instance, one usually reads and records to tenths only, no matter how precisely the rod may be graduated. (See Fig. 2.12.2.)

2.13 Differential Leveling

As previously mentioned, the universally useful reference plane for leveling is the invisible plane swept by a horizontal line of sight that is kept level by a spirit bubble. Using a spirit bubble connected to a sighting device, one can sight through a cross-hair telescope onto a graduated rod held vertical atop one point, and then read the rod held atop another point. Proper subtraction will give the difference of elevation between the two points. When differential leveling is being done, to set vertical control points, the routine is a sort of boot-strapping process. Starting at a known elevation, one sets up the instrument, reads a backsight on the rod held on the known bench mark, and then reads a foresight on the rod held on a point whose elevation is needed. Temporarily, this new

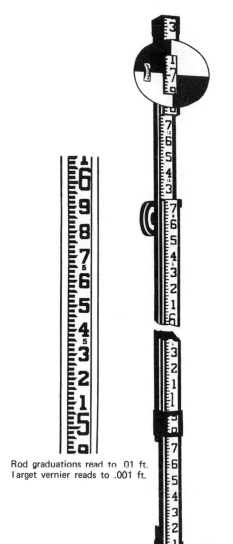

Rod graduations read to .01 ft.
Target vernier reads to .001 ft.

Fig. 2.12.1 Ordinary level rod.

point, a "turning" point, becomes the known elevation. The instrument is then moved forward and the process is begun anew to set a new point.

Thus, if the elevation of one point is known, then the elevation of the other point can be established for future use. Such known elevation points are called *bench marks,* permanent marks of known elevation. The routine is known as differential leveling.

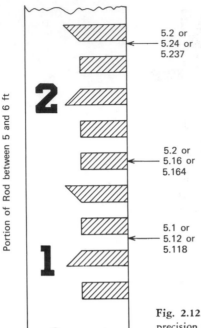

5.2 or
5.24 or
5.237

Portion of Rod between 5 and 6 ft

2

5.2 or
5.16 or
5.164

1

5.1 or
5.12 or
5.118

Fig. 2.12.2 Readings to various degrees of precision on ordinary level rod.

Comparing the notes (Fig. 2.13.1) and the diagram (Fig. 2.13.2) will permit an understanding of the process of differential leveling to set the new bench mark (BM 52) from the known bench mark (BM 51).

2.14   Leveling for Setting a Series of Bench Marks

When setting a bench mark (BM) or a series of bench marks for controlling construction, each must be used as a "turning point" for the level circuit. If a rod reading were taken on some point not directly involved in the level run, this would be a "side shot" and could not be regarded as reliable for elevation. This is simply because the elevation would not be carried through this side point, and it would not appear in the summation check; any blunder in reading would go undetected. Conversely, however, any point that has been used in the leveling circuit as a turning point could become a bench mark if it were needed and if its position were properly marked and safeguarded.

In setting a series of bench marks for vertical control, it is important always to "close the circuit" back to the starting bench mark or to another bench mark whose elevation is already known. This helps to

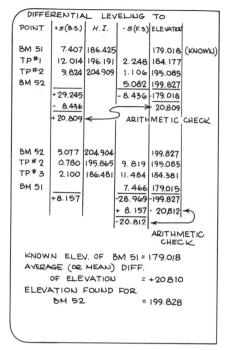
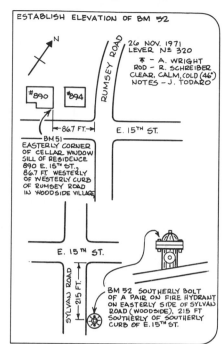

Fig. 2.13.1 Differential leveling note pages.

avoid incorporating unchecked blunders into the construction, with attendant costly consequences.

2.15   Relationships in the Engineer's Level

The care and adjustment of any engineer's level can be found in manufacturer's booklets and are not covered here. These are the basic relationships that should exist when the level is in adjustment.

Relationship *a*   The bubble tube must be at right angles to the vertical or azimuth or spindle axis. This assures that the bubble will remain centered no matter in which direction the telescope is pointed when spun about the vertical axis, once it has been set up.

Relationship *b*   The telescope alignment (line of collimation or line of sight) must be parallel to the bubble tube. This assures that the telescope line of sight is horizontal

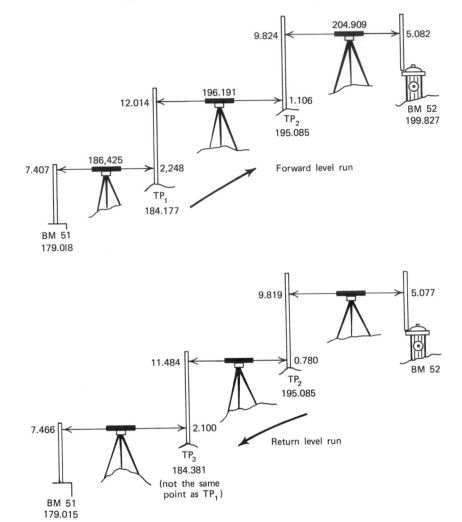

Fig. 2.13.2 Differential leveling.

when the bubble is centered, and is not tilted slightly upward or downward with respect to the bubble tube axis.

While relationship *b* is the more important of the two relationships, good practice while leveling dictates two precautions that will tend to compensate for any lack of these two adjustments.

Precaution *a*   At the instant of reading the rod, be sure the bubble is centered (i.e., immediately before and immediately after).

Precaution *b*   Keep pairs of sights equal in distance (i.e., have the rod on the foresight the same distance from the level as was the rod on the backsight).

These two precautions will help to avoid introducing bad errors into the leveling work.

2.16   Automatic or Self-Leveling Levels

Automatic levels, based on a pendulum or on a reflecting fluid surface, are now becoming common. These lend themselves well to construction work because of their ease of operation. One type is shown in Fig. 2.16.1.

Gravity actuates a pendulum "compensator," a suspended prism that ingeniously maintains a level line of sight. (See Fig. 2.16.2.) Or, in some cases, a pool of liquid reflects the ray of light in such manner as to accomplish the same result. Such "automatic" pendulum or gravity levels

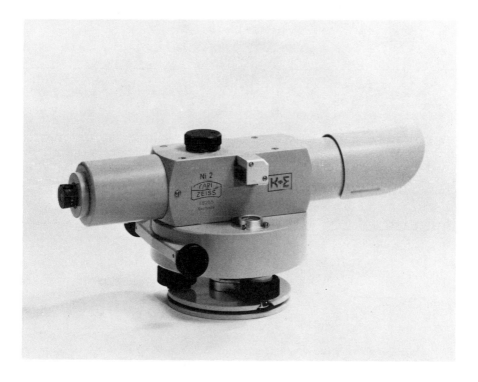

Fig. 2.16.1 Automatic or self-leveling level.   *Courtesy Keuffel & Esser Co.*

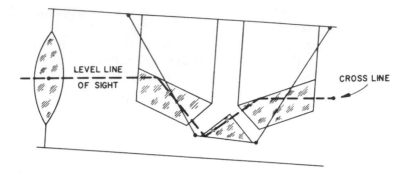

WHEN TELESCOPE TILTS UP
COMPENSATOR SWINGS BACKWARD

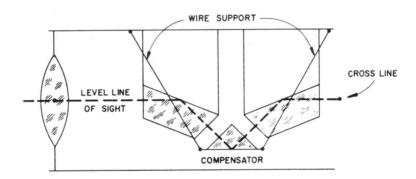

TELESCOPE HORIZONTAL

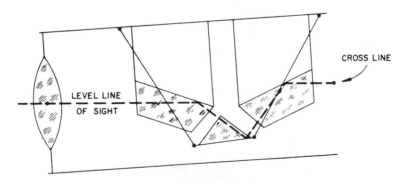

WHEN TELESCOPE TILTS DOWN
COMPENSATOR SWINGS FORWARD

Fig. 2.16.2 Compensator pendulum of automatic level.   *Courtesy Keuffel & Esser Co.*

are suitable for every type of construction or control leveling work. Once leveled by a small circular "bull's-eye" bubble, the compensating mechanism takes over and maintains a horizontal line of sight, largely insensitive to disturbing vibrations caused by wind or traffic or to changing climatic conditions. Its essential distinction is that the automatic level permits a considerably greater speed of observation than with conventional bubble levels.

2.17 Checking the Level

On construction the level must be checked for adjustment frequently, especially if doubt exists as to its accuracy. The procedure for adjusting to reestablish the correct relationships is not difficult, and is adequately described in various surveying books or instrument manuals. One must always be able to have confidence in his instrument.

An alert instrument operator will be able to set up a semipermanent level-adjustment arrangement, by having two firm points for which he finds the difference of elevation exactly by repeated measurements, or is able to set two scales S_1 and S_2, as in Fig. 2.17.1. He must do this from the central position A. Then, by moving his instrument to a position B he should get the same reading on S_1 as on S_2 (i.e., same difference of elevation as previously) or else he knows that an adjustment is needed. Compare this to Fig. 2.8.1 and refer also to Section 2.15.

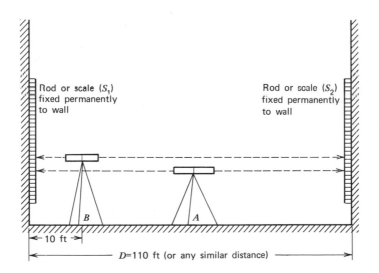

Fig. 2.17.1 Checking the engineers' level.

2.18   Leveling for a Profile

A profile of the ground is the trace of the ground on a vertical plane, a side view of the terrain. Profiles are usually drawn with the vertical scale exaggerated, as in the case of Fig. 2.18.1. By using a level to read the rod held at selected points along a given alignment, the elevation information (keyed to the distance along the line) can be found and used to plot the "profile" of the ground. It is greatly used in route work (highways, transmission lines, pipelines, etc.). (See Fig. 2.18.2.)

There are also airborne profile recorders that operate on radar, successful mainly in establishing the flying height of the airplane taking aerial photographs. More applicable in our context is a newly developing gas-laser profile recorder that can give a visual profile of considerable precision. (See Appendix B, on the laser.) With the laser recorder, tests indicate an airplane flying at about 1000 ft above the terrain can chart a profile to about 0.1 ft vertical accuracy. The laser is continuously beamed to the spot beneath the aircraft, and at the same time, continuous photography is made on which the laser line is superimposed, thus correlating the profile to visible objects below. As part of the routine, the plane overflies bench marks and other points of known elevation (road intersections, lakes, tennis courts, etc.) for intermittent check calibrations.

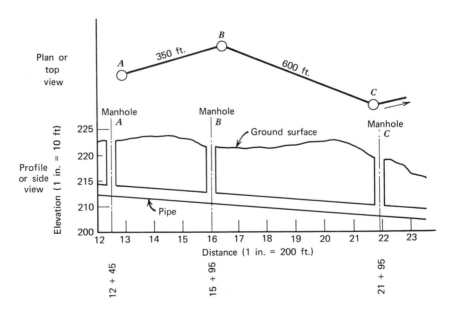

Fig. 2.18.1 Plan-profile of draininge pipe installation.

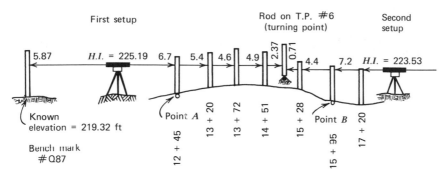

Fig. 2.18.2 How the ground profile is obtained by leveling along the route of drainage pipe.

The instrument shows promise as a means of extending vertical control adequately and inexpensively to remote construction sites. Profiling for power transmission lines, pipelines, microwave paths, highways, and so forth, will be practical applications as the method and equipment are further developed.

2.19   Stationing a Line

When taping a line, especially in highway or pipe-laying work, a handy method of marking or notation is to use 100-ft "length of station"; this means that Sta. 4 + 15.0 indicates a point that is 415.0 ft from the starting point (0 + 00.0), and so on. Stationing can tell a person at once how far from the beginning a particular mark is to be found or to be placed. The word "station" is understood as 100 ft, and we speak of "moving three stations" ahead, having "manhole *B* at Sta. 15 + 95," or "setting a transit at Sta. 17 + 00." Stationing is indicated in Figs. 2.18.1 and 2.18.2; it is evident that these illustrate only a portion of a longer drainage line. Figure 2.19.1 shows the field notes for the actual portion shown in Fig. 2.18.2, and the notes indicate both the rod reading and the stationing of the rod reading for each point. Locust Street is being extended east from manhole *A* at Elm Drive, and a drainage sewer is being planned beneath the street for some distance.

2.20   Leveling for a Cross Section

A cross section (*X*-section) is a profile made at right angles to a given alignment, for example, to a route alignment. Cross-section notes may be

STA.	+S(B.S.)	h.i.	-S(T.P.)	-S	ELEVATION
| PROFILE LEVELING FOR LOCUST STREET					
BM Q87	5.87	225.19			219.32
(A) 12+45				6.7	218.5
13+20				5.4	219.8
15+72				4.6	220.6
14+51				4.9	220.3
TP 6	0.71	223.53	2.37		222.82
15+28				4.4	219.1
(B) 15+95				7.2	216.3
17+20				7.7	215.8

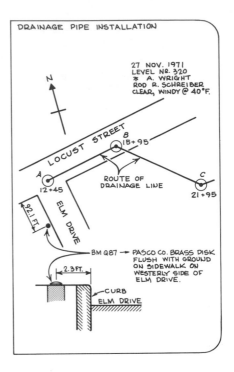

Fig. 2.19.1 Profile leveling notes.

done in a number of ways, so long as each elevation is keyed to its proper point. In Fig. 2.20.1, the stationing is shown to proceed up the page, a usual practice, so that a person holding the book and looking forward along in the direction of the line will be properly oriented as to "right" and "left." It is of interest that cross sectioning can sometimes be done very adequately with a hand level, because the results need not be better than the nearest tenth of a foot; if a more precise instrument be used, time need not be wasted in getting the more precise readings if there is no need for it. Figure 2.20.2 shows how the cross-section notes are obtained.

2.21   *Bench Marks at a Construction Site*

On a construction site for a structure, bridge, highway, and so on, it is important to place the project and its several components at the proper elevation. Control leveling will first be made to establish bench marks

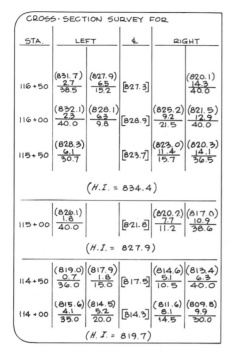

Fig. 2.20.1 Cross-section leveling notes.

nearby to assure this. Following that, many less permanent but more readily accessible bench marks of one sort or another will be set for convenience. Temporary bench marks for construction should be within 100 ft, at most 200 ft, of the location where they will be needed. Their location should be foreseen carefully enough to forestall any need to use a turning point between them and the constructed item (framework, pile cap, etc.) that needs an elevation check. Time and patience are often at a premium on construction jobs, so bench marks should be close enough to the work for quick access.

2.22 Semipermanent Instrument Setups

Controlling of elevations on a construction site can sometimes be made easier if a fixed position is maintained for the engineer's level. A concrete pier can perhaps be justified for a job of long duration, and the same instrument can be placed thereon each day. This has the advantage

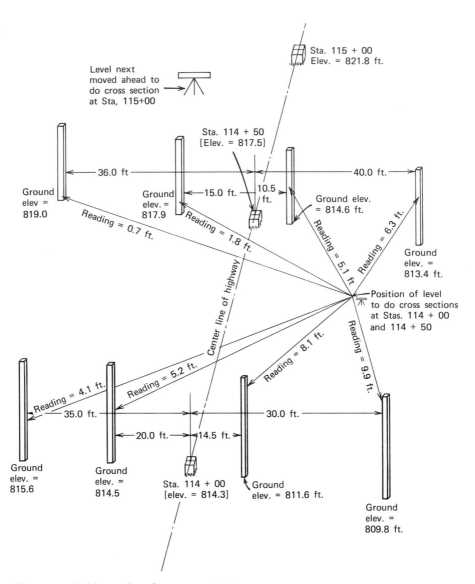

Fig. 2.20.2 Field procedure for cross sectioning.

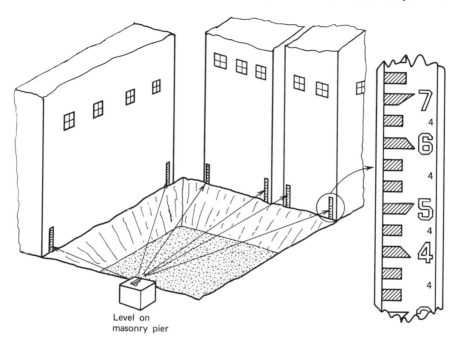

Fig. 2.22.1 Level setup to maintain a watch over building settlements during excavation.

that the height of instrument is unchanging, and the rod reading to any desired level is the same from day to day—a convenience for the workers.

There is reason sometimes also to arrange such a convenient permanent pedestal for the levelman who must watch for possible settlement of adjacent buildings during an excavation, especially one involving blasting or vibration. Such a situation is shown in Fig. 2.22.1. The instrumentman can sweep an array of targets set on the adjoining buildings and instantly see any settlement. This same procedure can be sometimes more conveniently done by using the same fixed-leg (not extension-leg) tripod and level instrument each day and setting the tripod shoes into notches cut into a sidewalk or pavement or rock. The level will always be at the same elevation, although care must be taken to set the same tripod up in the same notches each day. In any case, the instrumentman must assure himself frequently of the instrument's elevation by reference to one or more bench marks nearby, marks set back beyond the influence of the excavation work and vibration.

2.23   Cross-Section Leveling on a Gridiron Pattern

Sometimes the instrument must be down in the excavation, at the working level. Thus, the level is set up to give readings over a general area wherever needed, and the operator should again check frequently that his instrument is still correct by reference to one or more temporary bench marks. By using one or another setup, he can read the rod wherever it is held by his rodman and thus find the elevation of many points. He must refer to the bench mark often and should be sure the bubble is centered for each reading. This often is a situation where the readings might properly be made to tenths of a foot (instead of hundredths as here shown).

Notice in Fig. 2.23.1 that a grid system is used for conveniently identifying the points whose elevation is required periodically. This method is called the gridiron system and renders notekeeping very convenient, as seen in Fig. 2.23.2. To begin, the level is set conveniently where it can observe both the bench mark and the desired points on the gridiron. Sighting on a known point (BM 54) gives the elevation of the instrument (H.I. = 185.50), from which elevations of grid points can be found by sighting each point in turn. Care must be taken to avoid severely long sight distances, lest instrument maladjustment or even earth curvature introduce errors. Also, in these "side shots," blunders in reading cannot be detected, so extra care is needed.

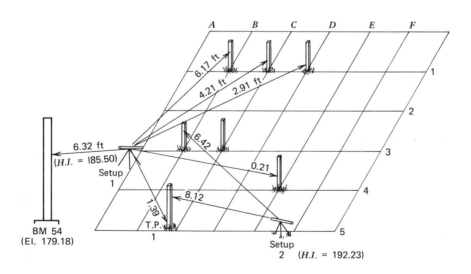

Fig. 2.23.1 Gridiron leveling (borrow-pit leveling).

POINT	+S	H.I.	-S(T.P.)	-S	ELEVATION
BM 54	6.32	185.50			179.18
A-0				9.43	176.07
A-1				7.85	177.65
A-2				5.19	180.31
A-3				3.26	182.24
A-4				1.37	184.13
A-5				0.41	185.09
B-0				6.99	178.51
B-1				6.17	179.33
B-2				3.03	182.47
C-0				6.42	179.02
C-1				4.21	181.29
C-2				1.73	183.77
C-3				0.36	185.14
D-0				8.35	178.15
D-1				2.91	182.59
E-0				2.08	183.42
E-2				1.19	184.31
E-4				0.21	185.29
TP #1	8.12	192.23	1.39		184.11
B-3				6.42	185.81
B-4				5.86	186.37
B-5				4.31	187.92
C-4				5.34	186.83

BORROW - PIT LEVELING, TRACY

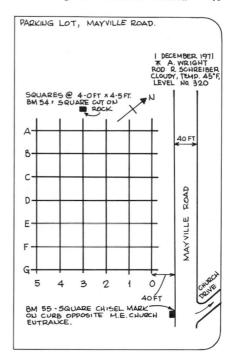

Fig. 2.23.2 Notes for gridiron or borrow-pit leveling.

The foregoing method may very well be used to take readings on a concrete floor slab or grade to detect local settlements, or to determine that it has been finished to specification. In such a case, reading the rod to thousandths of a foot would be necessary, and greater care would be taken. Perhaps, too, a more precise engineer's tilting level would then be selected for the work. This may also be the case in checking finished pavement grades for high-speed roadways or on viaducts, interchange ramps, or for certain factory floors where the finished floor level is critical.

In "gridiron cross-section" or "borrow-pit" leveling, as here indicated, the rodman must align himself fairly well in two directions by sighting pairs of ranging poles placed for this purpose. Or he can place himself by taping (or pacing) a proper distance along one range line. He might anchor the zero end of the tape to a zero stake, for instance. Sometimes it is more convenient to preset lath at the grid points in advance, especially if the terrain is very uneven. See Section 20.4, also.

The size of the grid square should be selected with judgment. While 50 by 50 ft squares could suffice for fairly uniform ground, 10 by 10 ft

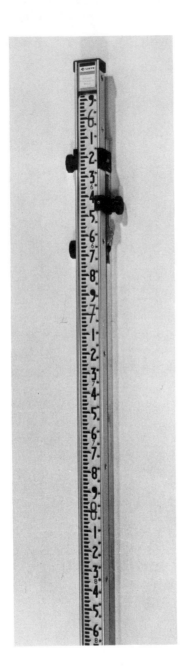

Fig. 2.24.1 Direct elevation rod.
Courtesy Lietz Co.

squares might be required for uneven or nonuniform terrain. Sometimes, too, extra points may be required, in the middle of a square, or along a range line, to assure the recording of some special feature. Care must also be used to insure the proper recording of each point, since it is easy to enter a reading in the wrong slot.

2.24   Rod for Direct Elevation Reading

A special direct rod exists for finding elevations rapidly, as in Fig. 2.24.1. The facing of the rod is a 10-ft adjustable continuous tape loop, which is set at a correct reading while being held on a known elevation; the loop is shifted until the rod reading corresponds with the known elevation of the point on which the rod is being held. For example, one will move the loop until it reads 9.61 if the rod is being held on a bench mark whose

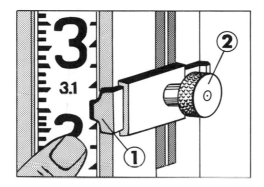

Fig. 2.24.2 Setting the tape loop at bench mark.

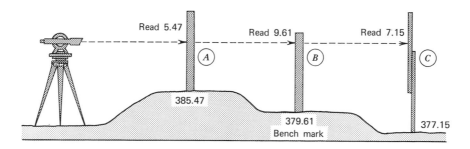

Fig. 2.24.3 Using a direct reading rod.

elevation is 379.61 ft; then the position of the tape face is clamped. (See Fig. 2.24.2.) As the rod is moved from point to point, because the face graduations progress from top to bottom, holding the rod on a higher point (e.g., 385.47) will give a higher rod reading (of 5.47); moving it, however, to a lower elevation (e.g., 377.15) will cause the rod to be read as 7.15 ft., and so on, as in Fig. 2.24.3. Its advantage is that it affords a direct elevation reading without the need to perform a subtraction to obtain each elevation.

3

Distance Measurements by Taping

3.1   Linear Measurement

Construction people use linear measurement all day long, reaching for a rule, a tape, a templet, or a measuring board to cut, to fit, or to set a component in place. A rule or a rod by being stiff can be held up, or out, or down; a tape is flexible, and needs two persons for handling, but it has its advantages, too. For short distances, rules work well, but for long measurements, tapes are called for, because they are quicker and more accurate.

Rules (such as the familiar folding rule of Fig. 3.1.1) are variously made and differently graduated, as can best be discovered by observing several types. Steel fabricators use steel rules, since the rule and the steelwork expand and contract alike with temperature changes, needing

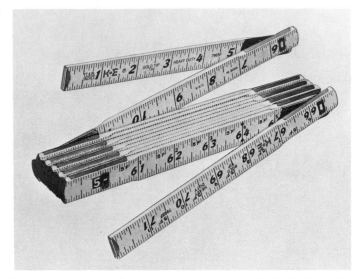

Fig. 3.1.1 Carpenter's folding rule.   *Courtesy Keuffel & Esser Co.*

no correction. Wood rules are commonly used in general construction. The foot, inch, and eighth-inch (or sixteenth-inch) are the usual graduations; machine-shop layout and the automobile industry ignore feet and work in inches and decimals of inches, a sort of "American metric system"; surveyors, in a somewhat similar manner, ignore inches and work in feet and decimals of a foot (tenths, hundredths, and thousandths). Rules are made in all these graduations—and in the universal metric graduations, too—and can cause blunders if inadvertently used wrong.

3.2   Feet versus Inches

A construction conversion is sometimes needed, from feet to inches (or vice versa), since engineers use feet and architects use inches. One simply multiplies by 12 (or divides by 12). For example, to make 21.81 ft usable for a carpenter, one can convert the 0.81 ft to inches:

$$0.81 \text{ ft} \times 12 = 9.72 \text{ in.}$$

$$21.81 \text{ ft} = 21 \text{ ft } 9.72 \text{ in.} = 21 \text{ ft } 9 \ 3/4 \text{ in., approx.}$$

or better, if needed, 21 ft 9 23/32 in.

On the contrary, to convert 8 ft 4 3/8 in. to feet, divide the inches by 12:

$$4 \ 3/8 \text{ in.} \div 12 = 0.3646 \text{ ft}$$

thus,

$$8 \text{ ft } 4 \, 3/8 \text{ in.} = 8.3646 \text{ ft}$$

or only 8.365 ft, or even 8.36 ft, depending on the amount of rounding-off desired.

Such simple converting can sometimes be a stumbling block for the unwary.

Another rule-of-thumb method is to know that 1/8 in. is just about equal to 0.01 ft. Thus, one can mentally make the conversions above:

$$21.81 \text{ ft} = 21.75 \text{ ft} + 0.06 \text{ ft} = 21 \text{ ft} + 9 \text{ in.} + 6/8 \text{ in.}$$
$$= 21 \text{ ft } 9 \, 3/4 \text{ in., approx.}$$

$$8 \text{ ft } 4 \, 3/8 \text{ in.} = 8 \text{ ft} + 3 \text{ in.} + 1 \, 3/8 \text{ in.} = 8.25 \text{ ft} + 11/8 \text{ in.}$$
$$= 8.25 \text{ ft} + 0.11 \text{ ft} = 8.36 \text{ ft, approx.}$$

Figure 3.2.1 shows this graphically.

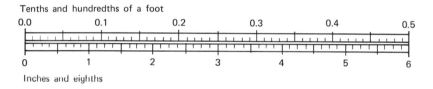

Fig. 3.2.1 Relation between eighths of an inch and hundredths of a foot.

3.3   *Cloth and Plastic Woven Tapes*

Tapes woven of strong synthetic yarns and covered by a flexible plastic coating are waterproof and light. Their ease of handling renders them extremely useful under many conditions where steel tapes would be impractical or cumbersome—and usable also with less danger around electrical lines. However, since all woven cloth or plastic tapes are liable to stretch or shrink slightly, they are not recommended where precise measurement is required. Using different pull values on a cloth or plastic tape while measuring between two points on the ground can best give one a sense of its stretchability. And comparing the cloth tape graduations with the steel tape will demonstrate how untrue the cloth tape can be. Cloth or plastic tapes are therefore graduated only to tenths of feet, giving some warning that one should not expect an accurate result. (See Fig. 3.3.1.)

Fig. 3.3.1 Woven tape.   *Courtesy Keuffel & Esser Co.*

3.4 Steel Tapes

Steel tapes are accurate, light but strong, and resistant to wear and abrasion. They are rugged and reliable, but they can rust unless cleaned and lightly oiled after use. Steel tapes are usually marked with graduations to hundredths of a foot, and readings can be estimated to thousandths of a foot. (See Figs. 3.4.1 and 3.4.2. Tapes made of steel are more difficult to use than rules. (See Fig. 3.4.3.) They sag, and thus can give erroneous end readings; pulling harder to eliminate the sag error can then actually stretch them elastically, as rubber stretches; and heat or cold can introduce appreciable temperature change in a long tape. So, knowing something about tape errors is necessary if accuracy of a fairly high order is required. The error formulas are discussed later, along with some examples to show the size of some errors.

Steel tapes are not intended to be exactly correct when lying flat

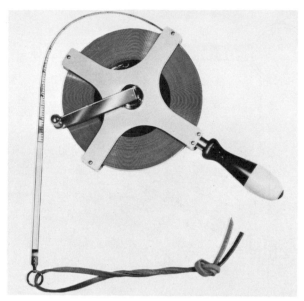

Fig. 3.4.1 Steel tape.   *Courtesy Keuffel & Esser Co.*

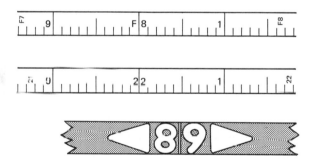

Fig. 3.4.2 Styles of steel tape graduations.

(fully supported) under zero tension. Most are manufactured to read true length with some specific value of pull, say about 10 lb. The best way to be sure about any given tape is to try it against a known standard of length, like a 100-ft distance accurately marked out for this purpose in a basement corridor. Or it may be compared with a tape standardized by the U. S. Bureau of Standards for the manufacturer at the time of purchase.

Steel tapes, therefore, are usually manufactured to be correct length

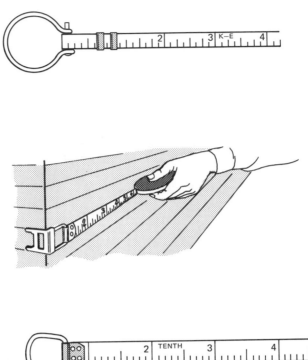

Fig. 3.4.3 Styles of end fasteners on steel tapes.

only when (a) fully supported throughout, (b) pulled with a specified tension, such as 10 lb, and (c) used at a temperature of 68° F. These three conditions must be present simultaneously in order to obtain correct results. If the tape is lifted off the ground and held at the ends only, we have another situation. An amount of sag shortening will be introduced by removing the support throughout its length—such as by holding the tape a few feet above the ground. When it is supported at the ends only, it will sag and the ends come closer together; pulling with a greater tension will stretch the tape (as well as lessen the sag), and these two effects will combine and will tend to correct the sag error. It is quite possible that a pull value can be discovered that will correctly compensate for the sag and give a correct length. This will be discussed in Section 3.5.3.

If the tape is fully supported throughout its length and then the tension is increased or decreased, the tape will expand or shrink, giving a *pull error*; this error can be calculated, should the need arise. If the temperature falls below 68°F, the tape will shrink; if it rises above 68°F, the tape will expand. A *temperature error* can thus occur. The following examples give some illustration of the problem of sag, pull, and temperature errors in taping.

3.5 Errors in Steel Tapes

Taping errors are introduced here with examples to give a practical means of handling steel tapes with the confidence of knowing how large or how small they may be under different circumstances.

3.5.1 Pull or Tension Error

$$E_P = \frac{PL}{AE}$$

is the formula for pull error, where
P = pull (lb)
L = length (ft)
A = cross-sectional area of tape (sq in.)
E = modulus of elasticity (psi) (for steel, E = 30,000,000)

Let us assume that Tape No. 1016 is a lightweight 100-ft steel tape, ¼ × 0.02 in. in cross section, which is of correct length under a 10-lb tension fully supported and is now stretched out on the ground and held with a 15-lb tension, 5 lb more than standard. It will stretch this amount (giving a pull error) beyond its correct value:

$$E_P = \frac{5 \times 100}{0.005 \times 30,000,000} = 0.0033 \text{ ft, or } 0.003 \text{ ft}$$

A pull of 20 lb (10 above the standard) will bring this error to twice this amount, or 0.0067 ft, or 0.007 ft. A pull of 30 lb (20 above standard) will bring the error to 0.0133 ft, or 0.013 ft. A tension handle or tensiometer should be used to measure the value of the pull applied by the tapemen. (See Table 3.5.3.1, p. 56.) Such a tension handle is shown in Fig. 3.5.1.1.

Fig. 3.5.1.1 Tape tension handle.   *Courtesy Keuffel & Esser Co.*

3.5.2   Sag Error

$$E_s = \frac{W^2L}{24P^2}$$

is the formula for sag, where

W = weight between supports (lb)

L = length (ft)

P = pull (lb)

Assume that No. 1016, the same tape as used above, (since steel weighs 490 lb/ft^3) weighs

$$W = \frac{0.02}{12} \times \frac{0.25}{12} \times 100 \times 490 = 1.70 \text{ lb}$$

The computed sag error (for a pull of 10 lb) is:

$$E_s = \frac{1.70^2 \times 100}{24 \times 10^2} = 0.121 \text{ ft}$$

As an alternative to computing the weight, the tape can be weighed. An approximate but good method to weigh it is to first weigh the tape and reel intact; then remove the tape and weigh the reel alone. The difference in the two weights will be the weight of the 100-ft tape (W) to be placed in the formula—nearly enough.

3.5.3 *Combined Sag and Pull Error*

Thus it is apparent that if this tape is correct when pulled at 10-lb tension, fully supported throughout its length, and then the full support is removed and the tape becomes end-supported only under a 10-lb tension, it will sag; the ends will come closer together by 0.121 ft and the tape ends will be only 99.879 ft apart. The tape does not really become shorter; it only means that the end marks (0 and 100) are drawn closer together by some amount. A greater tension than 10 lb will be required to restore the end marks to their original position. So, this sag error can be lessened by increasing the tension. For example, if the pull is increased to 20 lb, the sag error will then be:

$$E_s = \frac{1.70^2 \times 100}{24 \times 20^2} = 0.0303 \text{ ft}$$

considerably less than that for 10 lb. The sag error for a 30-lb tension would be:

$$E_s = \frac{1.70^2 \times 100}{24 \times 30^2} = 0.0133 \text{ ft}$$

somewhat less than that for 20 lb.

It would appear impossible to increase the tension enough to take the sag error completely out of the tape, but remember that by increasing the tension one also stretches the tape elastically. This is very helpful. There can exist, then, a specific pull value that will cause the sag shortening to compensate exactly for the lengthening due to pull; it is called the "normal tension" for the tape, the tension at which sag error is cancelled by pull error.

Calculating some values for E_p and E_s of Tape No. 1016 and plotting the two curves illustrates how the pull error can compensate the sag error for a particular value of tension. The values are shown in Table 3.5.3.1.

These values are shown by the curves plotted as Fig. 3.5.3.1. The "normal tension" (P_N) occurs where the curves cross, where the sag error is balanced by the tension error.

Table 3.5.3.1   *Combined Pull and Sag Errors of Tape No. 1016*

Tension	E_p	E_s	Combined E	Distance Between End Marks
0	(?)	∞	∞	0.000 (why?)
10	0.000[a]	0.120	−0.120	99.880
15	0.0033	0.0532	−0.050	99.950
20	0.0067	0.0300	−0.023	99.977
25	0.0100	0.0192	−0.009	99.991
30	0.0133	0.0133	−0.000	100.000
35	0.0167	0.0098	+0.007	100.007

[a] The 10-lb pull can be thought of as standard in the present instance. When $P = 10$ lb, the elastic stretch in the tape is exactly the amount that was present when the tape was initially standardized, so the E_p can be properly written as zero. However, removing the full support throughout will introduce a sag error of 0.120 ft, and E_s at 10 lb is not zero.

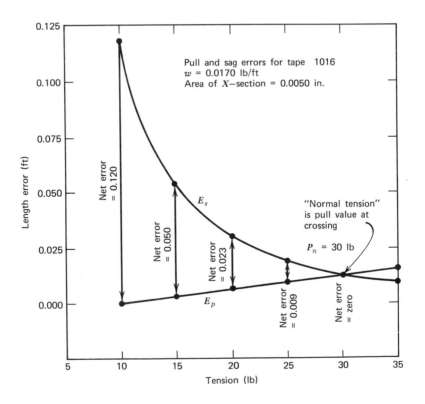

Fig. 3.5.3.1 Plot of errors to show "normal tension."

Thus by computation it is found that by pulling on the end-supported tape—this particular tape—with a 30-lb pull, the result is the same as if the tape were pulled with 10 lb when fully supported throughout its 100-ft length. As a practical matter, by trial-and-error comparison with a standard, one can discover a correct tension for each of two conditions: (1) tape fully supported throughout its length, and (2) tape end-supported only. Then one's particular tape can be used confidently either way. Large sag errors can then be avoided by taking care to apply proper tension when holding the tape at the ends and using plumb bobs to lay out or measure distance. No effort is made here to discuss the errors for partial lengths (e.g., 78.16 ft, 39.21 ft, etc.). Generally these errors will be rather negligible in the shorter lengths, and the preceding discussion will be something of a guide for taping lengths less than 100 ft.

3.5.4 Temperature Error

The formula for temperature error is

$$E_t = k \cdot L \cdot \Delta t,$$

where

k = thermal coefficient of expansion (for steel, k = 0.00000645)

L = length (ft)

Δt = difference in temperature from standard (°F)

If the previous tape (No. 1016) is used at 40°F, but is correct in length at 68°F, the temperature error per 100 ft is E_t = 0.00000645 × 100 × 28 = 0.018 ft. Temperature error is caused by the expansion of the tape when heated or the contraction of the tape when cooled. For example, to lay out 467.25 ft on the ground with a steel tape at 40°F, one must first lay out the 467.25 ft and then add a correction amount (= 4.67 × 0.018 = 0.08). Setting out 467.33 ft (apparent) would thus give a value on the ground of 467.25 ft (true). By reverse reasoning, if a ground distance were reported as 467.33 ft at 40°F, the true distance would be found by subtracting the 0.08 ft, to get 467.25 ft. In general, a 15°F difference gives about a 0.01-ft change in a 100-ft steel tape (actually 0.0097 ft change per 100 ft).

The thermal coefficient for steel is 0.00000645 ft per ft per °F. An alloy called invar is also used for tapes, having a thermal coefficient of 0.0000002 (1/30 of steel), but such tapes are used only for very high-precision work. Invar is a nickel-steel alloy, quite costly and more readily damaged in use.

Steel tapes are manufactured so as to be correct in length at 68°F (20°C). Serious errors can be introduced in taping if temperature variation is not considered, frequently much greater errors than those of sag. Thus when long distances are laid out or checked at low or at high temperatures, the temperature correction must be considered.

3.5.5 Gradient or Slope Error

When a tape is not held level, the resultant slope can introduce errors in layout or measurement of distance. Care must be exercised, since it is not easy to sense when the tape is inclined. To be sure, a tape level can be used; a hand level sighted at the other tapeman is the most frequently employed device. The slope is often given as a percent grade; for example, a 5% grade means $\Delta h = 5$ ft when $L = 100$ ft. Percent grade is very commonly used to describe gradient or slope. See Fig. 3.5.5.1. When a measurement is made along an incline, if the gradient is known or found (as, with a hand level), the gradient-error formula is used.

Tape error due to *gradient*:

$$E_g = \frac{\Delta h^2}{2L}$$

where

Δh = difference in height of ends of tape (ft)

L = length (ft)

A measurement on a slope will not give a horizontal distance unless this error is properly applied as a correction. If a distance is measured on a 2.5% slope (i.e., 2.5-ft rise in 100 ft), the correction to be applied for each 100 ft is:

$$E_g = \frac{2.5^2}{2 \times 100} = 0.031 \text{ ft}$$

Thus if a slope distance was reported as 240.00 ft, the true (horizontal) distance is:

$$D = 240.00 - (0.031)(2.40)$$
$$= 240.00 - 0.07 = 239.93 \text{ ft}$$

Similarly, if a 240.00-ft horizontal distance is to be laid out along a slope of 2.5%, the actual distance marked off on the slope should be:

$$240.00 + 0.07 = 240.07 \text{ ft}$$

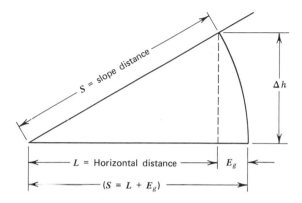

Fig. 3.5.5.1 Relationship between horizontal and slope distances.

The foregoing tape errors are given simply to indicate their order of magnitude and their effect in ordinary work, and also to indicate how to cope with them if the need should arise. It frequently occurs that more careful and precise measurements should be made in construction than are made. On the other hand, when ordinary construction measurements are made, knowing that sizable errors can arise from neglect of tension, or temperature, or slope, and so on, one will be careful not to insist that his measurements are exact. And, whenever circumstances warrant, one might judge that an approximate answer will suffice, and it may be possible then to ignore these error sources.

3.6 *How to Handle Taping Errors Practically*

Probably the best way to do ordinary taping well is to learn, for a given steel tape, the proper tension to make it read exactly 100 ft when stretched between two correctly calibrated end marks. Then one may simply use it under that tension in everyday operations and make proper allowances for temperature only; all other corrections (except gradient) will be properly cared for without any calculation. This may require that a standard base be set up by use of a standard tape, and that an occasional recheck be made thereon of the working tape.

There will be two values of pull to be employed: (*a*) the value for the tape when fully supported throughout; and (*b*) the value when the tape is end-supported only. When taping along the ground, as in Fig. 3.6.1, one uses the first tension value, applying the correction for temperature

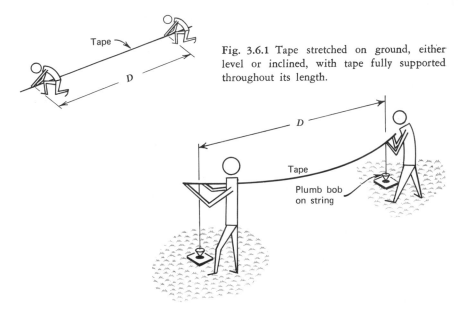

Fig. 3.6.1 Tape stretched on ground, either level or inclined, with tape fully supported throughout its length.

Fig. 3.6.2 "Plumbing down" with the tape held at 0 and 100 marks, end-supported only.

(and for slope if the ground is not level). When "plumbing down" at the ends of the tape held horizontal as in Fig. 3.6.2, one uses the second tension value, applying the correction for temperature (but not for slope). And a thermometer can be used to check for any necessary temperature adjustment.

3.7   Ordinary Taping Procedures

If a tape is known not to be exactly correct in length, but if the value of the error is known, the tape can still be very easily used. If it is, say, too long by 0.032 ft per 100 ft, one can lay out a 200-ft building by measuring off 200.000 ft with the tape, then backing up a distance of $2(0.032) = 0.064$ ft. In effect, then, the distance set out is correct, but would seem to be $200.000 - 0.064 = 199.936$ ft. To lay out 200 ft if the tape is too short by 0.032 ft per 100 ft, one would lay out the apparent 200.000 ft and move ahead a distance of 0.064 ft. In effect, the distance is correct, but would, if measured with that tape, seem to be $200.000 + 0.064 = 200.064$ ft.

As a handy way to remember, think first of *laying out* a distance. The *measuring* of a distance between two points with a tape of known error is handled oppositely to find the correct distance. A simple table is given here as a memory aid.

When Tape Is	To Lay Out a Distance	To Measure a Line
Too long	Subtract	Add
Too short	Add	Subtract

Taping with a steel tape graduated to hundredths is sometimes simply intended to find feet and tenths (87.6), sometimes feet and hundredths (87.57), and sometimes feet and thousandths (87.568). It depends on the precision demanded for the particular task. If a result to the nearest foot or half-foot or tenths is wanted, a cloth tape can probably be used. In other cases, a steel tape ought to be employed. When the thousandths of feet is required, it is obvious that the last (thousandths) digit must be achieved by estimation between the hundredths marks, and a steel tape must be used, quite carefully attending to any error sources.

3.8 Special Taping Procedures

Often a 300-ft tape can be used to advantage instead of a 100-ft tape, saving time without sacrificing accuracy. This is especially true where the taping is being done on level or uniformly sloping ground and with no traffic problems. It is apparent that a 300-ft tape cannot be held end-supported only by two tapemen; such tapes are almost always fully supported along the ground. If the ground is uniformly level, a horizontal distance directly results; if the ground is uniformly sloping, it is necessary to apply a proper slope correction to obtain a horizontal distance.

Where tapemen cannot work along the ground, some improvising may be needed. In flat marsh country, sometimes marsh buggies or helicopters can be used to tow the tape along, in which case it is common to use a 1000-ft cable. In this case the cable rides along the marsh grass and is essentially supported throughout its length. Also, each thousand-foot station is marked by a 2 in. × 2 in. × 14 ft "lath" driven into the muck. To guard against a dropped 1000-ft station on the ground, the laths are premarked and used in numerical sequence.

Sometimes where street or highway traffic is a problem, taping may have to be done in short odd-length segments to cross a busy thorough-

Fig. 3.8.1 Measuring wheel.

fare. Here particular care must be exercised to record and add all the random lengths correctly, without omitting any.

For some less accurate types of measurement, a rolling measuring wheel affords a quick method (Fig. 3.8.1). Pushed along the ground, it records distance, although (unfortunately) not true horizontal distance. Attached to the 4-ft circumference wheel are an automatic tally register and an easy reset arrangement. Where lengths of pavement or curbing need to be measured, where roadside obstruction information is to be collected, or wherever "reasonably close" distances will serve, the wheel is good. To get information for planning road-widening or curb-and-gutter improvement, or waterline or sewerline planning, it could serve admirably and do the job at less cost.

Alternative means of measuring distance are available, and will be covered in Chapter 5. The methods discussed are stadia, subtense, tacheometry, range-finding, and electronic distance measurement. These are indirect methods, depending on the evaluation of some other quantity that gives a measure of the desired distance. Many are essentially less accurate than the taping procedure, although often quite sufficient for many purposes; others are as accurate as or more accurate than taping.

4

Angle Measurements and Directions

4.1   The Transit and Theodolite

A transit or theodolite is basically an angle-measuring instrument. It measures horizontal angles in the horizontal plane about an azimuth (vertical) axis, and measures vertical angles in a vertical plane about an elevation (horizontal) axis. It is equipped with spirit bubble tubes for establishing the horizontal and vertical planes. As may be seen by studying Figs. 4.1.1 and 4.1.2, the transit is a theodolite. The word "transit" came into being in the United States simply by reason of the ability of the telescope to turn 180° about the horizontal (elevation) axis, and sight in the opposite direction. Usage in the United States has imparted to "theodolite" the notion of a more precise instrument, and also the notion

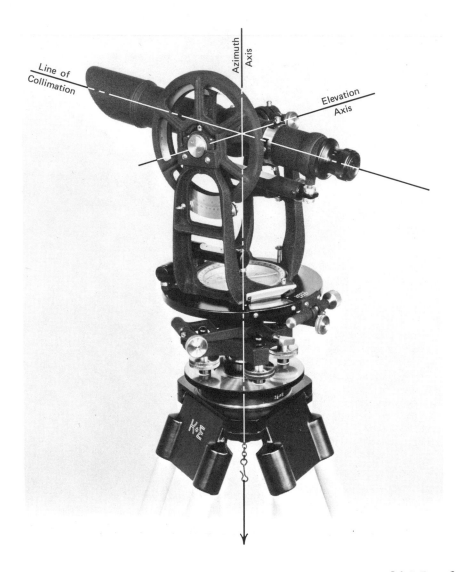

Fig. 4.1.1 Transit.

Courtesy Keuffel & Esser Co.

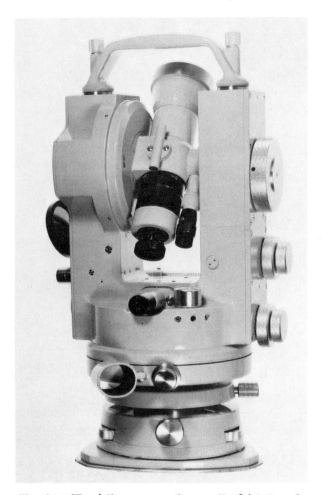

Fig. 4.1.2 Theodolite.   *Courtesy Keuffel & Esser Co.*

that it reads directions only and cannot readily measure or set off horizontal angles. The two words are getting to be rather interchangeable now, since the instruments are.

4.2   The Telescope Cross-Wire Reticule

Transit or theodolite "line of sight" is fixed by the telescope axis, really by the optical center of the lens system and a set of cross wires mounted in the telescope. The cross wires or cross hairs are called the reticule; some patterns are shown in Fig. 4.2.1. The vertical cross-wire marks the

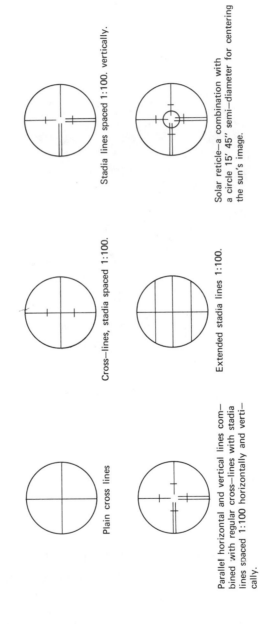

Plain cross lines

Parallel horizontal and vertical lines com—
bined with regular cross—lines with stadia
lines spaced 1:100 horizontally and verti—
cally.

Cross—lines, stadia spaced 1:100.

Extended stadia lines 1:100.

Stadia lines spaced 1:100. vertically.

Solar reticle—a combination with
a circle 15′ 45″ semi—diameter for centering
the sun's image.

Fig. 4.2.1 Telescope reticule patterns in common use.

Fig. 4.2.2 Transit telescope sighting mark on bridge centerline. *Courtesy Kern Instruments.*

alignment for the instrumentman, either to sight a fixed mark, or to set a mark on line, or to read a horizontal angle. The single vertical wire can "split" a rod or pencil or nail held vertically over a point, or be made to coincide with the center of a target. The double vertical wire can be used to bracket a plumb-bob cord, or a distant rod or pencil or nail. The horizontal cross wire usually marks a level or elevation, either to read a level rod to discover the elevation of a point, or to set a stake or mark at a desired elevation, or to read a vertical angle.

4.3   The Plumb Bob or Plummet

The transit is centered over a ground point or hub by means of a brass plumb bob or plummet suspended on a string from the center of the instrument. (See Fig. 4.3.1.) Thus correct horizontal position of the transit can be established for measuring horizontal angles.

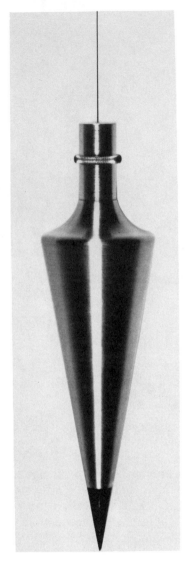

Fig. 4.3.1 Plumb bob or plummet.   *Courtesy Kern Instruments.*

The optical plummet is a device for centering a transit or theodolite over a point, a replacement for the plumb bob hanging beneath the instrument. By sighting horizontally through a right-angle prism as in Fig. 4.3.2, one can see the point over which the theodolite is centered when the instrument is leveled. It serves the same purpose as the plumb bob on a string but is not affected by the wind. It assures vertical alignment

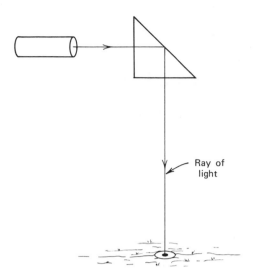

Ray of
light

Fig. 4.3.2 Principle of optical plummet.

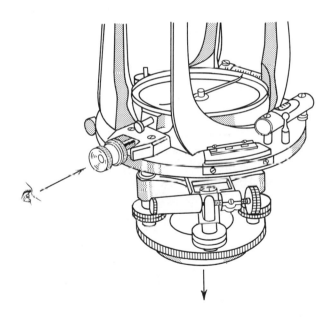

Fig. 4.3.3 Optical plummet on typical theodolite.

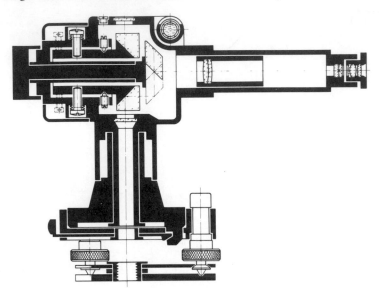

Fig. 4.3.4 Zenith and nadir plummet.   *Courtesy Wild Heerbrugg.*

as accurate or more accurate than that furnished by the plumb bob. A transit so equipped is shown in Fig. 4.3.3.

Sometimes a theodolite equipped with an optical plummet is mounted on a bracket or structure to provide a vertical line for a shaft or a building or a tower. The ordinary optical plummet is quite adequate for maintaining verticality, even for a moderately high structure or for a moderately deep shaft. Commercially available optical plummets give an accuracy of about ± 0.02 in. in 100 ft.

Generally, the accuracy of the ordinary optical plummet is less than may be needed to drop horizontal alignment to the bottom of a deep shaft; wires may then have to be used for aligning a very deep tunnel. However, special optical plummets are available that fit on a tripod or other support and supply vertical alignment for much greater vertical heights. Centered over a ground point, such an optical plummet, as in Fig. 4.3.4, can be aimed upward to maintain the vertical for high-rise construction of any type. Pointed downward into a shaft, it can be used for maintaining vertical alignment during construction without the difficulty, the time, and the uncertainty of suspending long piano wires and heavy weights. This matter will be covered again to some extent in Chapter 16.

4.4 Tripods

In the United States a usual tripod head is the four-screw leveling head, although the use of the three-screw leveling head is being accepted more and more by American construction and surveying people, since it is standard on imported instruments. The three-screw leveling head can cause the instrument to change in elevation if it be releveled during a leveling task; this is a feature of its construction, and one ought to be aware of it. For most other work, there is no cause for concern, and the three-screw head is easier to use.

An innovative feature of some American manufacturers is the parallel-shift tripod, seen in Fig. 4.4.1, which makes lateral sliding of the tripod to move the transit over a point a lot easier. Otherwise, to set over a point and level the instrument at the same time is a bit difficult, although it is not impossible for a beginner or bothersome to a practiced hand.

PARALLEL-SHIFT TRIPOD

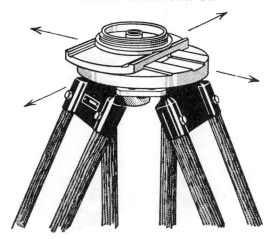

Fig. 4.4.1 Parallel-shift tripod.   *Courtesy Keuffel & Esser Co.*

4.5 Automatic Centering

European manufacturers have made their theodolites so they attach to the tripod through a tribrach ("star-plate") or other adapter that can detach from the theodolite and remain on the tripod. It is then possible for a target to be placed exactly in the same position, exactly centered—forcibly, automatically—over the point where the instrument was used. (See Fig. 4.5.1.) Thus, in fact, the target can be sighted from a new posi-

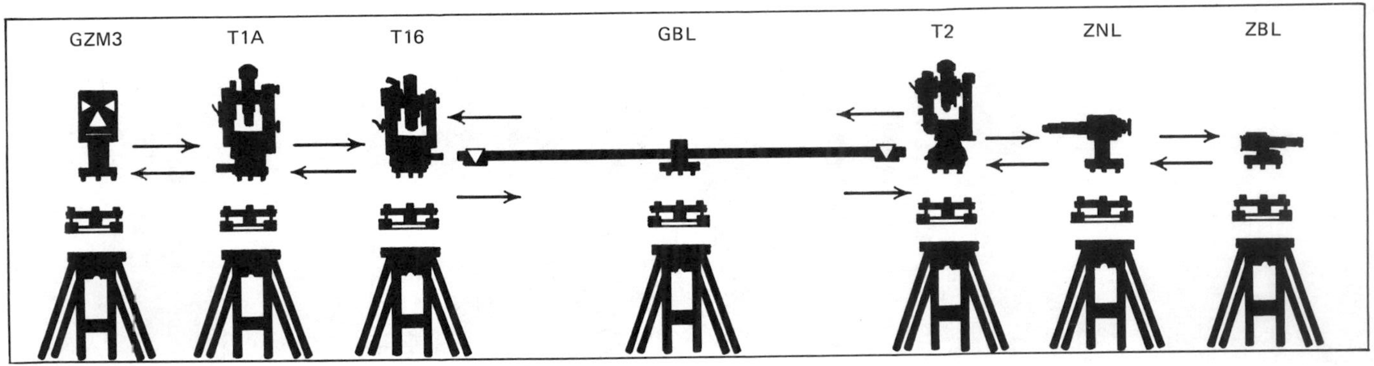

72

Fig. 4.5.1 Automatic centering.

GZM3   T1A   T16   GBL   T2   ZNL   ZBL

Courtesy Wild Heerbrugg.

tion with great exactness, and with the assurance that there is no eccentricity by reason of one's not placing the target directly over the point. The opposite situation is also possible, placing the theodolite over the point where the target was. This method is of benefit in precise traversing work or in triangulation. Essentially, then, the tripod stays put until all the work over the point is accomplished, and the several devices are placed on it alternately in sequence with confidence that they each in turn are automatically centered over the point.

4.6 Relationships in the Transit or Theodolite

The necessary relationships that must exist in the transit or theodolite are given here. Refer to instrument operations manuals for the method of verifying that these relationships exist and adjusting the instrument so they will.

Relationship *a*   Plate bubble should center when azimuth axis is vertical (in direction of gravity).

Relationship *b*   Vertical X-wire should lie in a plane perpendicular to the elevation axis.

Relationship *c*   The line of sight should be perpendicular to the elevation axis.

Relationship *d*   The elevation axis should be perpendicular to the azimuth axis.

Relationship *e*   The telescope bubble should center when the line of sight is horizontal.

Relationship *f*   The vertical circle should read zero when the line of sight is perpendicular to the azimuth axis.

4.7 Universal Surveying Instrument

Although essentially regarded as a device to measure angles, the transit (or theodolite) is the universal tool for alignment. By rotating the telescope about its horizontal (elevation) axis, one can sweep a vertical plane; by rotating about its vertical (azimuth) axis, one can sweep a horizontal plane—much as the engineer's level does. By combining these motions, the transit can be pointed virtually in any direction to sight or set points on a line. The following are some very common and most important functions the instrument serves.

4.8 Prolonging a Straight Line

The transit (theodolite) frequently must be used to extend a line AB, to a point C. There are three distinct cases:

Case 1. Occupying A with the transit, to prolong the line past B to C, there usually is little difficulty or danger of error. No angles are measured; one merely sights and marks points. (See Fig. 4.8.1.) A pencil or plumb bob held on B gives the first sight, from which one easily sets the marks at C by observing a pencil or plumb bob held at that point. If points B and C are at quite different elevations, and thus if there is an appreciably different telescope inclination, a check might be made by repeating the procedure, this time with the telescope inverted. If C falls at a different point, the mean of the two markings should be used. This would indicate that the elevation axis of the transit is not normal to its azimuth axis and ought to be adjusted.

Case 2. Occupying B with the transit, to prolong the line AB past B to C, a "transiting" of the telescope (inverting it on its elevation axis) is called for. (See Fig. 4.8.2.) The transit is backsighted to a pencil or bob at A, the telescope is transited or "plunged," and a point is marked at C. (This is shown, for our illustration, as C_1 on the sketch.) Then, with telescope still inverted, the transit is again backsighted to A, transited or plunged once more (back to normal or erect), and a second point is set at C (shown as C_2 on the sketch). These points C_1 and C_2 ought to coincide; if not, the correct point C is midway between them.

Case 3. Straddling-in on a straight line is another version of "prolonging" a straight line, in a sense. To "straddle in" on line AB (or "buck in") is to set up the transit on C (between A and B) without being able to occupy either A or B with the instrument. The procedure begins with an estimate or guess, and a temporary setup is made at C_1 and the transit is leveled. (See Fig. 4.8.3.) A sight is made on point A, and the telescope is inverted (plunged) to see if point B is on the line AC. Usually it is not, so a new point C_2 is chosen by estimation, and the process is repeated: sighting of A, inverting to sight B. If B is not yet on line, the process is repeated as often as need be.

Finally, when C seems to be exactly on line AB, the double-sighting

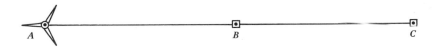

Fig. 4.8.1 Prolonging a line, case 1.

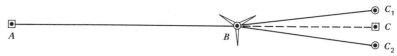

Fig. 4.8.2 Prolonging a line, case 2.

Fig. 4.8.3 Straddling-in on a straight line, case 3.

procedure of Case 2 is used to refine the work. If double-sighting now shows that the point C is not exactly on line, a slight shifting of the instrument head will usually suffice to put it on line—and a final check should again be made before point C is set. When it is, the point can be marked directly beneath the plummet.

This double-sighting procedure, once with telescope direct and once with telescope inverted, is an error-compensating practice that should always be used. It is also called "double-centering." It ensures that instrumental errors will not creep into the construction layout.

4.9 Using the Transit for Plumbing a Wall or Column

If a mark high up on a structure is to be placed by a transit sighting in direct alignment with a point on the ground, there may be an appreciable elevation angle. It is needful, therefore, to sight the point below A and mark the point above twice, once with the telescope direct and once with the telescope inverted. The mean of the two marks set will be the proper location of the point, as seen in Fig. 4.9.1.

The reason that points B_1 and B_2 may not coincide is that the elevation axis of the transit may be slightly out of adjustment, that is, not perpendicular to the azimuth axis. It is important, of course, before attempting to set the point B, to be sure that the azimuth axis of the transit or theodolite be truly vertical. The reverse process, that is, setting a lower point A in alignment with a higher point B, is accomplished in much the same way.

In the foregoing, a little practice with his theodolite or transit and a

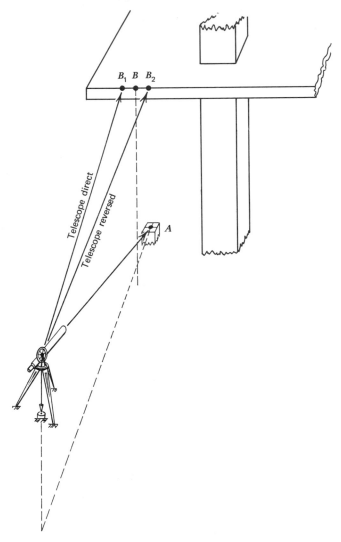

Fig. 4.9.1 Placing an alignment mark high on a structure.

consideration of the degree of accuracy required by the work will influence the instrumentman's decision in these matters. The placing of a
mark on line high up on a wall, or verifying that a high point on a building is aligned with one below, or generally sweeping a vertically aligned
plane is mostly a transit or theodolite function. One encounters some
difficulty and may need to employ an eyepiece prism to sight very steeply

upward, and it is impossible with a transit to sight vertically or nearly vertically downward. The optical plummet, described in Section 4.3, may serve in such cases.

4.10   The Automatic Level as a Plumbing Device

There is, however, a little known and fairly recent device that can be employed in these vertical or nearly vertical plumbing situations, an adaptation of the automatic level described in Section 2.16. Attaching a right-angle rotatable prism to the telescope converts the instrument to a mechanism capable of very exact plumbing upward or downward. The 90° prism clamped in front of the objective lens deflects the true horizontal line of sight into a true vertical plane. Then as the prism is rotated about the telescope axis to sweep the vertical plane, it can be used to sight or set a point anywhere in that vertical plane. By repeated meas-

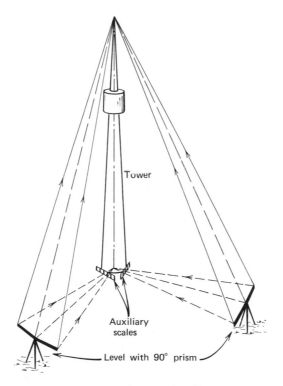

Fig. 4.10.1 Automatic level with prism used to plumb radio tower.

urements, an accuracy of less than 0.2 sec of the vertical can be achieved for checking high-rise buildings, television towers, tunnel shafts, and so on. Figure 4.10.1 suggests its use in checking the verticality of a radio mast by plumbing it from two directions. This instrument is described also in Section 16.7 in connection with assuring verticality of a shaft.

4.11   Reading the Circles of Transit or Theodolite

Transits read to 1 min, to 30 sec, to 20 sec, or even to 10 sec of arc. The finer precisions are the more costly transits, and most construction measurements can be done with the less expensive instruments. The American transit has circles made of bronze, brass, or aluminum, sometimes coated with silver for engraving the graduations and the numbers. Verniers are used to help read the fine graduations of these circles, as seen in Fig. 4.11.1; the vernier is a mechanical device to subdivide the scale. Examination of a vernier will show how coincidence of lines determines the reading; the eye can readily detect scale divisions that are in alignment.

The metal circles of the American transit are difficult to read when finely divided, since there is a distinct limit to the magnification one can employ to read the light reflected from the metal circle's surface. Thus, the transit usually will read to 1 min, 30 sec, or 20 sec—or half these values by estimation, when vernier lines do not exactly coincide but two adjacent ones nearly do.

With graduations scribed on glass circles through which the light passes, however, an "optical" theodolite or "optical" transit is able to read much finer graduations through use of much greater magnification possible in a lens system. Hence it is fairly easy to read such a theodolite to 0.1 min directly (which is 6 sec), or with the aid of an optical micrometer to 1 sec or even a fraction of a second. The use of glass circles orginated in Europe, and hence we generally call such instruments "European" theodolites—properly, "optical-reading" instruments.

Interestingly, while one must read the two opposite verniers separately and average them for accuracy on the American transit, the use of light rays passing through glass circles allows the two opposite sides of the circle to appear in one image in the optical-reading theodolite. Also, instead of the vernier, the optical micrometer measures the distance on the circle from the nearest circle graduation, and this value also appears in the image in the viewing window with the main circle reading. In Fig. 4.11.2, the path of the light is seen to travel through both vertical and horizontal circles, and through opposite sides of each to pick up the imagery of numbers and graduations etched on the glass. Figure 4.11.3

GRADUATED 30 MINUTES READING TO ONE MINUTE
DOUBLE DIRECT VERNIER

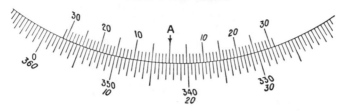

This is an ordinary double direct vernier, reading from the center to either extreme division (30). The circle is graduated to half degrees, and the vernier (from 0 to 30) comprises 30 divisions; consequently, the value of one division on the vernier is 30 minutes ÷ 30 = 1 min.

The figure reads 17° + 25′ from left to right and 342° 30′ + 05′ = 342° 35′ from right to left.

GRADUATED 20 MINUTES READING TO 30 SECONDS
DOUBLE DIRECT VERNIER

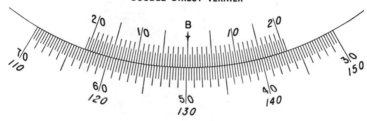

This is also a double vernier, reading from the center to either extreme division (20). The circle is graduated to 20 min or 1200 sec and there are 40 divisions in the vernier; consequently, the value of one division on the vernier is 1200 sec ÷ 40 = 30 sec.

The figure reads 130° 00′ + 9′ 30″ = 130° 9′ 30″ from left to right, and 49° 40′ + 10′ 30″ = 49° 50′ 30″ from right to left.

GRADUATED 15 MINUTES READING TO 20 SECONDS
DOUBLE DIRECT VERNIER

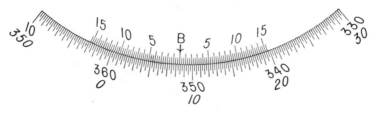

This is a double direct vernier, reading from the center to either extreme division (15). The circle is graduated to 15 min or 900 sec and there are 45 divisions in the vernier; consequently, the value of one division on the vernier is 900 sec ÷ 45 = 20 sec.

The figure reads 8° 15′ + 9′ 20″ = 8° 24′ 20″ from left to right and 351° 30′ + 5° 40″ = 351° 35′ 40″ from right to left.

Fig. 4.11.1 Styles of transit verniers.   *Courtesy Keuffel & Esser Co.*

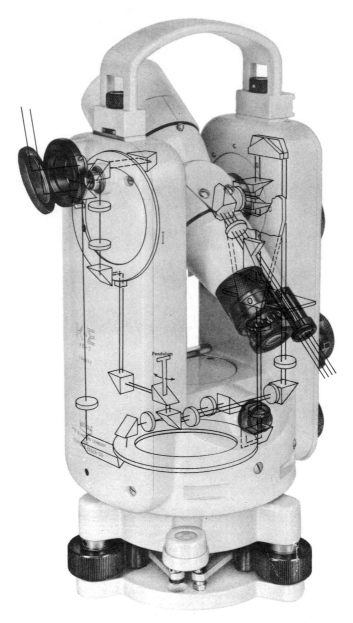

Fig. 4.11.2 Path of light rays in an optical-reading theodolite showing how opposite sides of the circles are simultaneously read and averaged. *Courtesy Keuffel & Esser Co.*

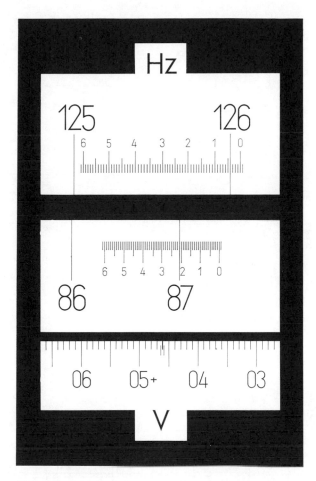

Fig. 4.11.3 Reading of one type of optical theodolite.

shows a reading of both horizontal and vertical circles; Fig. 4.11.4 shows two different readings of a slightly different system that employs an optical micrometer that requires the operator to make the main circle reading coincide with the double index line. In the small auxiliary window of Figs. 4.11.4 and 4.11.5 is seen the fine reading; this is the angular distance that the optical micrometer (Fig. 4.11.6) moves along the circle to achieve coincidence. There are several variations of optical reading systems, none very difficult to master. Instruction manuals give sufficient information, and assistance is readily supplied by each manufacturer.

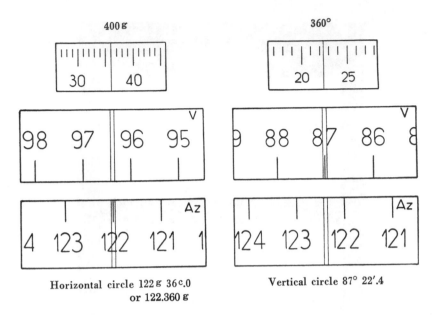

Horizontal circle 122ᵍ 36ᶜ.0
or 122.360ᵍ

Vertical circle 87° 22'.4

Fig. 4.11.4 Readings of a 6-sec optical theodolite, with 360° and 400ᵍ systems illustrated.
Courtesy Wild Heerbrugg.

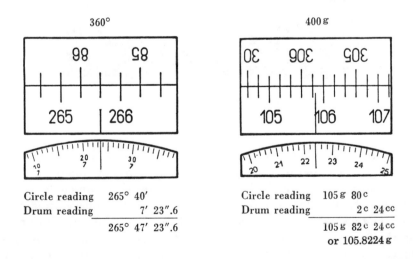

Circle reading   265° 40'
Drum reading   7' 23".6
————————————————
265° 47' 23".6

Circle reading   105ᵍ 80ᶜ
Drum reading   2ᶜ 24ᶜᶜ
————————————————
105ᵍ 82ᶜ 24ᶜᶜ
or 105.8224ᵍ

Fig. 4.11.5 Reading of a 1-sec optical theodolite, with 360° and 400ᵍ systems illustrated.
Courtesy Wild Heerbrugg.

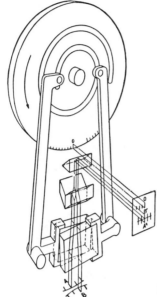

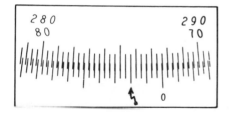

Fig. 4.11.6 Diagram of the optical micrometer.

Fig. 4.11.7 Vernier reading of transit equipped with glass circles. *Courtesy Leitz Co.*

Interestingly, Fig. 4.11.7 illustrates a vernier reading of an optical theodolite. The use of back-lighted glass circles enables the superposition shown, which makes the reading easier.

4.12   Horizontal Angles with Transit or Theodolite

In using the transit to measure a horizontal angle or to set out a horizontal angle, the initial sighting is normally with the index set at 0°00′00″, as shown in Fig. 4.12.1. However, any random initial reading can be used. Had the initial position reading of the angle in the sketch been 102°51′ and the final position reading been 174°34′, a subtraction of

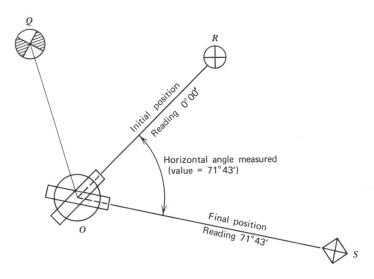

Fig. 4.12.1 Top view of transit measuring horizontal angle.

the two values would have also given the value of the angle as 71°43'. It is simply a matter of one's convenience to set initially at zero.

The American transit has two indexes, A and B, set 180° apart so that opposite sides of the horizontal circle can be read and averaged for greater precision. If index A (and its vernier) were to read 39°51'30" on one side, the other index, B (and its vernier), should read 219°51'30" on the opposite side. Generally they will be in good accord, but any eccentricity of the circle resulting from wear or usage will be eliminated in effect by using the average of the A and B readings.

For instance, if angle QOR of Fig. 4.12.1 is being measured and the A and B vernier readings are not in agreement, the mean of the two can be counted on to give a reliable value.

Point Occ.	Point Obs.	Reading	A	B	Mean
O	Q	0°45'	30"	00"	15"
	R	58°32'	40"	20"	30"
		Angle = 57°47'15"			

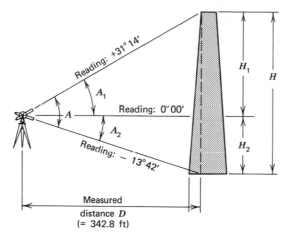

Fig. 4.13.1 Height of chimney stack by vertical angles.

4.13 Vertical Angles with Transit or Theodolite

The vertical circle is used to measure an angle of elevation about the elevation (horizontal) axis. The angle is either plus (if inclined upward) or minus (if inclined downward). Review Section 1.6 for zenith angles, however.

To obtain the height of a chimney stack (H) of Fig. 4.13.1 for instance, we must add H_1 and H_2. Angle A would not be really useful; fortunately the angle does show as $+A_1$ and $-A_2$. Then the computation can be done:

$$H = H_1 + H_2$$
$$H = D \cdot \tan A_1 + D \cdot \tan A_2$$
$$H = 342.8 \,(0.60642) + 342.8 \,(0.23377)$$
$$= 207.88 + 83.56$$
$$= 291.44 \text{ ft}$$

The elevation of a mountain peak, for example, might be found by measuring the vertical angle from one or more points of known elevation. (See Fig. 4.13.2.)

Incidentally, the height of a pole or building can also be roughly measured without a transit. By measuring its shadow (L_B), as also the shadow

(L_P) of a 10-ft rod held vertical, the pole or building height can be found by proportion, as in Fig. 4.13.3. It is of interest that on vertical aerial photographs the heights of towers, tanks, buildings, and poles are often thus measured by photo-interpreters using shadow measurements from a single photograph on which the height of some one object casting a shadow is known or can be estimated.

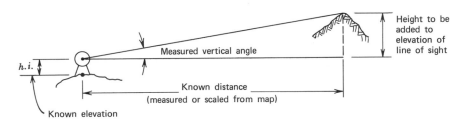

Fig. 4.13.2 Finding height of mountain by vertical angle.

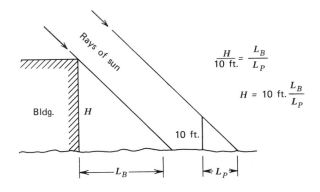

$$\frac{H}{10 \text{ ft.}} = \frac{L_B}{L_P}$$

$$H = 10 \text{ ft.} \frac{L_B}{L_P}$$

Fig. 4.13.3 Rough height measurement by proportional of lengths of sun's shadows.

5

Distance Measurements by Other Methods

5.1 *Distance Measurement by Stadia*

Stadia (a Latin word implying stride—about 6 ft, or double the "pace")
is a term applied today to a distance measurement with a theodolite or
transit telescope. Cross wires, optically imposed on the field of view, are
seen to intercept a graduated rod held plumb at some distance away.
From the proportionality of similar triangles in Fig. 5.1.1, it is seen that
the distance is the sum of $d + (f + c)$; d is a multiple of S, usually 100
times the value. The $(f + c)$ constant is whatever value is fixed by the
manufacturer of the telescope, and is usually about 1 ft. If S is 2.85, and

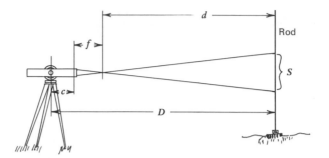

Fig. 5.1.1 Distance measurement by stadia with telescope horizontal $(f + c \neq 0)$.

$(f + c)$ is 1 ft, assuming this instrument's stadia multiple is 100, the total distance to the rod in this case can easily be determined as

$$D = 100 \ (2.85) + 1 = 286 \text{ ft}$$

Modern telescopes are internally focusing, sealed against dust, and are now usually built so that $(f + c) = 0$. This reduces the equation of distance to

$$D = kS$$

where k is some convenient value, like 100. If the $(f + c) = 0$, then the present distance would be only 285 ft. Figure 5.1.2 shows two different distances measured by the stadia method where the $(f + c) = 0$.

This simple case, with horizontal (or level) line of sight, gives the distance without trouble, although its precision depends upon the operator's ability to read the rod graduations. Usually the precision obtained is about ± 1 ft in 250 ft, though careful and repeated measurements can

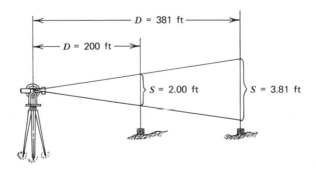

Fig. 5.1.2 Distance measurement by stadia with telescope horizontal $(f + c = 0)$.

give twice that precision, or 1/500. At best one must regard the stadia method as approximate and not sufficiently reliable for setting out stakes for construction. It serves well to gather distance information for plotting maps or for other work that does not require high precision. Its obvious advantage is that many more distances can be measured and points located in much less time and with much less effort than by taping.

5.2 Inclined Stadia Sights

The stadia method can be used also with inclined sights, utilizing vertical angle information furnished by the transit at the time of sighting the rod. Horizontal distances can be ascertained by formula, as can vertical distances (differences of elevation). From Fig. 5.2.1 one can readily see the derivation of the needed formulas, atlhough a simple circular "stadia slide rule" will give both these vertical and horizontal distance values quite painlessly and within the accuracy of an admittedly approximate method.

It can be seen that

$$S' = S \cos \alpha$$
$$DIST = d + (f + c) = 100 \, S' + (f + c)$$
$$= 100 \, S \cos \alpha + (f + c)$$
$$H = DIST \cos \alpha$$
$$= 100 \, S \cos^2 \alpha + (f + c) \cos \alpha$$
$$V = DIST \sin \alpha$$
$$= 100 \, S \cos \alpha \sin \alpha + (f + c) \sin \alpha$$

If $(f + c) = 0$, the formulas reduce to:

$$H = 100 \, S \cos^2 \alpha$$
$$V = 100 \, S \cos \alpha \sin \alpha$$

These formulas can be solved, or a stadia slide rule can be used, or a set of tables can be employed to give the H and V values. Note an important feature; H will always be *less* than $DIST$ (the inclined or slope distance), but that V can be either *plus* or (if the telescope is inclined downward) *minus*. And if the telescope is sighted directly at the value on the rod which is the *h.i.* (i.e., the height of telescope above instrument point), the value V is also exactly the difference in elevation between the stake sighted and the stake over which the instrument is set, as shown in Fig. 5.2.2.

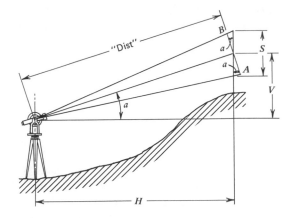

Fig. 5.2.1 Distance measurement by stadia with telescope inclined.

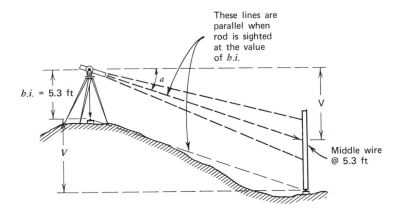

Fig. 5.2.2 The "*V*" is the difference of elevation between the ground points.

5.3 The Stadia Procedure in the Field

To use the stadia method, the transitman can usually keep the notes and draw the sketches if one rodman is employed, or he may need a note-keeper if two or more rodmen are used. As a quick procedure, the transitman sights the rod, putting the middle X-wire at the "*h.i.*" value, and then at once moves the lower wire to the nearest full foot mark. Reading the distance between bottom and top wires as his intercept (*i*), he notes it as "distance" in the record. Then he resets the middle wire on the *h.i.*, and waves "OK" to the rodman, who is free then to move to a new point. While the rodman is on the move, the instrumentman reads and records both vertical and horizontal angles. Final reduction of notes to find H and V is done later at leisure by tables (Fig. 5.3.1) or stadia reducer (Fig. 5.3.2).

For long stadia sights, special wide-faced rods with bold markings 3 in. wide are available. (See Fig. 5.3.3.) For short-sight stadia work, a level rod can be used. Stadia measurements can be done rapidly by an experienced field crew and are sufficiently accurate for topographic mapping, for building-site informational surveys, and whenever great accuracy is not desired.

Stadia with a three-wire level is a procedure that should be mentioned here. Achieving a distance measurement by use of a rod intercept is used

TABLE A
HORIZONTAL CORRECTIONS FOR
STADIA INTERCEPT 1.00 FT.

Vert. Angle	Hor. Cor. for 1.00 ft.	Vert. Angle	Hor. Cor. for 1.00 ft.	Vert. Angle	Hor. Cor. for 1.00 ft.
0°00'	0.0 ft.	5°36'	1.0 ft.	8°02'	2.0 ft.
1°17'	0.1 ft.	5°53'	1.1 ft.	8°14'	2.1 ft.
2°13'	0.2 ft.	6°09'	1.2 ft.	8°26'	2.2 ft.
2°52'	0.3 ft.	6°25'	1.3 ft.	8°38'	2.3 ft.
3°23'	0.4 ft.	6°40'	1.4 ft.	8°49'	2.4 ft.
3°51'	0.5 ft.	6°55'	1.5 ft.	9°00'	2.5 ft.
4°15'	0.6 ft.	7°09'	1.6 ft.	9°11'	2.6 ft.
4°37'	0.7 ft.	7°23'	1.7 ft.	9°22'	2.7 ft.
4°58'	0.8 ft.	7°36'	1.8 ft.	9°33'	2.8 ft.
5°17'	0.9 ft.	7°49'	1.9 ft.	9°43'	2.9 ft.
5°36'		8°02'		9°53'	3.0 ft.
				10°03'	

Results from Table A are correct to the nearest foot at 1000 feet and to the nearest 1/10 foot at 100 feet, etc.

With a slide rule, multiply the stadia intercept by the tabular value and subtract the product from the horizontal distance.

Example. Vertical angle, 4°22'; stadia intercept, 3.58 ft.
Corrected Hor. Dist. =
358 − (3.58 × 0.6) = 356 ft.

Table B gives the vertical heights for a stadia intercept of 1.00 ft. With a slide rule, multiply the stadia intercept by the tabular value.

Example. Vertical angle, 4°22'; stadia intercept, 3.58 ft.
Vertical Height = 3.58 × 7.59 = 27.2 ft.

TABLE B

VERTICAL HEIGHTS FOR STADIA INTERCEPT 1.00'

Min.	0°	1°	2°	3°	4°	5°	6°	7°	8°	9°
0	0.00	1.74	3.49	5.23	6.96	8.68	10.40	12.10	13.78	15.45
2	0.06	1.80	3.55	5.28	7.02	8.74	10.45	12.15	13.84	15.51
4	0.12	1.86	3.60	5.34	7.07	8.80	10.51	12.21	13.89	15.56
6	0.17	1.92	3.66	5.40	7.13	8.85	10.57	12.27	13.95	15.62
8	0.23	1.98	3.72	5.46	7.19	8.91	10.62	12.32	14.01	15.67
10	0.29	2.04	3.78	5.52	7.25	8.97	10.68	12.38	14.06	15.73
12	0.35	2.09	3.84	5.57	7.30	9.03	10.74	12.43	14.12	15.78
14	0.41	2.15	3.89	5.63	7.36	9.08	10.79	12.49	14.17	15.84
16	0.47	2.21	3.95	5.69	7.42	9.14	10.85	12.55	14.23	15.89
18	0.52	2.27	4.01	5.75	7.48	9.20	10.91	12.60	14.28	15.95
20	0.58	2.33	4.07	5.80	7.53	9.25	10.96	12.66	14.34	16.00
22	0.64	2.38	4.13	5.86	7.59	9.31	11.02	12.72	14.40	16.06
24	0.70	2.44	4.18	5.92	7.65	9.37	11.08	12.77	14.45	16.11
26	0.76	2.50	4.24	5.98	7.71	9.43	11.13	12.83	14.51	16.17
28	0.81	2.56	4.30	6.04	7.76	9.48	11.19	12.88	14.56	16.22
30	0.87	2.62	4.36	6.09	7.82	9.54	11.25	12.94	14.62	16.28
32	0.93	2.67	4.42	6.15	7.88	9.60	11.30	13.00	14.67	16.33
34	0.99	2.73	4.47	6.21	7.94	9.65	11.36	13.05	14.73	16.39
36	1.05	2.79	4.53	6.27	7.99	9.71	11.42	13.11	14.79	16.44
38	1.11	2.85	4.59	6.32	8.05	9.77	11.47	13.17	14.84	16.50
40	1.16	2.91	4.65	6.38	8.11	9.83	11.53	13.22	14.90	16.55
42	1.22	2.97	4.71	6.44	8.17	9.88	11.59	13.28	14.95	16.61
44	1.28	3.02	4.76	6.50	8.22	9.94	11.64	13.33	15.01	16.66
46	1.34	3.08	4.82	6.56	8.28	10.00	11.70	13.39	15.06	16.72
48	1.40	3.14	4.88	6.61	8.34	10.05	11.76	13.45	15.12	16.77
50	1.45	3.20	4.94	6.67	8.40	10.11	11.81	13.50	15.17	16.83
52	1.51	3.26	4.99	6.73	8.45	10.17	11.87	13.56	15.23	16.88
54	1.57	3.31	5.05	6.79	8.51	10.22	11.93	13.61	15.28	16.94
56	1.63	3.37	5.11	6.84	8.57	10.28	11.98	13.67	15.34	16.99
58	1.69	3.43	5.17	6.90	8.63	10.34	12.04	13.73	15.40	17.05
60	1.74	3.49	5.23	6.96	8.68	10.40	12.10	13.78	15.45	17.10

Fig. 5.3.1 A typical set of stadia reduction tables.   *Courtesy Keuffel & Esser Co.*

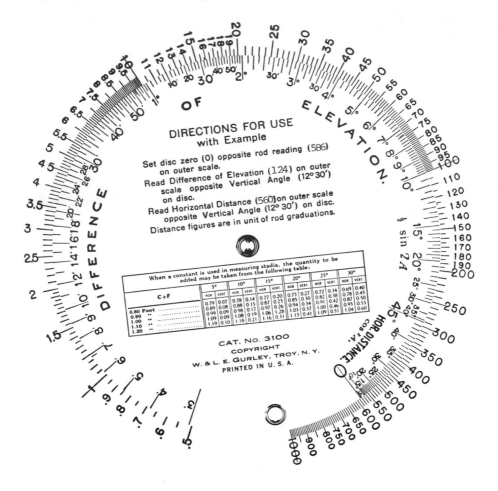

Fig. 5.3.2 Cox's stadia computer.   *Courtesy Teledyne Gurley.*

in three-wire differential leveling, a fairly recent method in the United States. Using an engineer's level equipped with three horizontal wires, all three wires are read on the rod and averaged for a more precise reading. The difference between top and bottom wire gives a stadia measure of distance, and the level operator can be guided by these readings in keeping his foresight distances equal to his backsight distances. This suggests that, in fact, a level can in certain flat terrain be used to obtain topographic information if it is equipped with a three-wire reticule and a circle to read horizontal directions. (See Appendix C.)

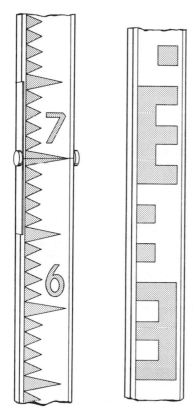

Fig. 5.3.3 Two types of stadia rods marked so as to be visible at greater distances.

5.4 *Distance by the Subtense Bar*

A distance-measuring method using a theodolite and a subtense bar (Fig. 5.4.1) has considerable value under difficult circumstances as a substitute for taping. For distances between 100 and 250 ft its accuracy lies between 1:12,000 and 1:5000. For greater distances, one might use a succession of short segments.

Subtense distance measurement is done by measuring a horizontal angle between two marks on a horizontal bar some distance away. The horizontal angle is the angle "subtended" by (or included between) the marks set (usually) 2 m apart on the subtense bar.

This subtense method usually requires a theodolite reading to 1 sec (although a transit could be pressed into use) because normally an in-

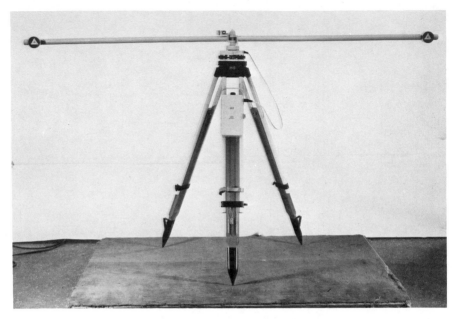

Fig. 5.4.1 Subtense bar on tripod; it can be folded for easy carrying.

Courtesy Keuffel & Esser Co.

accurate angle measurement renders the distance too uncertain to be any good. The 2-m bar is of invar and is accurate to about five digits (2.0000 m), so the angle should be accurate to five digits for consistency. If not, the distance is that much less accurate. The distance D is found from simple trigonometry, as seen in Fig. 5.4.2:

$$D \text{ (in meters)} = \frac{1.0000 \ m}{\tan \beta/2} = 1.0000 \ m \ \text{cotan} \ \beta/2$$

Then, to convert meters to feet, multiply by 3.28084; a simpler method is to read the distance in feet from a table of distances supplied by the manufacturer of the subtense bar.

Very reliable distances can be obtained by the subtense method, especially in the short ranges from 20 to 200 ft, provided a precise instrument (theodolite, preferably, reading to 01″ of arc) is used. Applications that suggest themselves for this procedure are measurement across a busy thoroughfare, around high-tension lines or third rails, across gullies, chasms and bodies of water, or wherever taping is difficult, as on steep

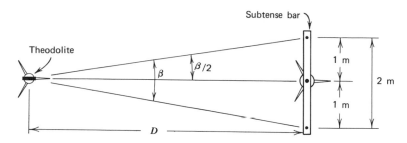

Fig. 5.4.2 Top view of distance measurement by subtense.

slopes. With regard to this last mentioned use, there is an advantage in that the distance D as found by subtense is the correct horizontal distance, not the inclined or slope distance. No conversion to horizontal is needed. The reason is simply that one measures the horizontal angle between the marks on the bar, and that angle would be the same wherever the theodolite may be located vertically at the point where it measures the angle.

5.5 *Subtense Triangulation*

From an understanding of subtense measurement, one might reason to the use of a bar longer than 2 m, say a "bar" of 300 ft laid out by taping or by subtense, and setting targets for sighting across a river. The principle is applicable; one ought to be careful to keep the ratio of distance to "subtense bar" less than 20.1, for fictitious accuracies may otherwise result.

In Fig. 5.5.1, line AB must be set perpendicular to CD, and the distance AB must be measured very carefully with a tape or with a subtense bar, since length AB itself in this case serves as a "subtense bar" for the cross-river measurement. Distance AB should be at least 1/20 of the distance CD for acceptable results. The angle β is measured with a theodolite to seconds. Then,

$$CD \text{ (in ft)} = \frac{\frac{1}{2} AB \text{ (in ft)}}{\tan \beta} = \frac{1}{2} AB \cotan \beta$$

This procedure is very similar to "triangulation," so it is called "subtense triangulation" sometimes, or "trig-traverse."

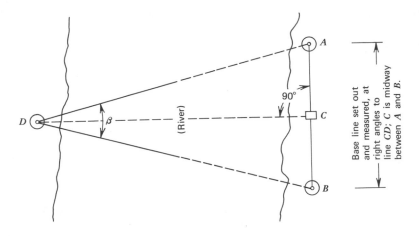

Fig. 5.5.1 Subtense principle for measuring long distance.

5.6 *Distance by Tacheometer*

Other variations of such distance-measuring equipment are in use, some of which are here indicated. Optical tacheometry is European in origin. The tacheometer is a stadia theodolite having a cross-hair distance that varies with the inclination of the telescope. This simplifies stadia distance measurement by keeping the multiplying factor constant. No matter what the rod intercept, the horizontal distance to the rod is, say, 100 times the intercept. And a simple factor appears simultaneously in the field of view that permits a rapid mental calculation of the difference of elevation to the point where the rod is held.

Distance measurement by any type of "tacheometer" is not very difficult, and the instrument gives the difference of elevation as well. It is only a matter of getting a little practice with the particular instrument. Americans have not heretofore used such instruments extensively, though it can be anticipated that they will be using them more in the near future. They are an extension of the "optical theodolite" and should find ready acceptance here.

In Fig. 5.6.1, for example, the horizontal distance to the rod is given by the intercept between the outer pair of curved lines: the difference of elevation, by the intercept between the inner pair. The rod pattern is unusual, but one can readily get accustomed to it. The multiplication constants, 20, 50, or 100, or in other instruments $+0.1$, $+0.2$, $+\frac{1}{2}$, or $+1$ when the telescope aims up and -0.1, -0.2, $-\frac{1}{2}$, or -1 when the telescope

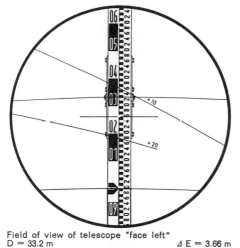

Field of view of telescope "face left"
D = 33.2 m   Δ E = 3.66 m

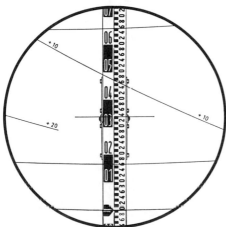

Field of view of telescope "face right"
D = 32.9 m   Δ E = 4.71 m

Reading of Horizontal Distance and Difference of Elevation

The telescope contains two reticules. The fixed one contains two vertical lines, a cross mark and the stadia lines with a ratio of 1:100. On the movable reticule are etched four slightly curved diagram lines. The difference in elevation is given by the intercept between the inner pair of lines, and the horizontal distance by the intercept between the outer pair. The rod readings are taken alongside the vertical lines of the fixed reticule. The difference of elevation thus obtained refers to the rod reading at the cross mark. The constants to be used for measuring difference of elevation are marked on the reticule as follows:

constant 20 = ||
constant 50 = |||||
constant 100 = |

Fig. 5.6.1 The tacheometer. Curves for distance *and* elevation permit direct determination of horizontal distance *and* difference of elevation even in the case of inclined sights. *Courtesy Carl Zeiss.*

aims down, are constants that appear in the field of view and can be applied by mental calculation to the intercept.

The tacheometer can give distance and elevation for cross sectioning for highway work in rough terrain, where difference of elevation by use of an engineer's level would require much moving of the instrument. By sending rodmen out to each side of the route center line, a tacheometer operator can get elevations and distances from fewer setups rather quickly.

5.7   *Distance by Range Finder*

The artillery or camera range-finder is somewhat the reverse of the subtense bar, having its fixed base mounted on the instrument itself, with two lines of sight bringing the images into coincidence on the target sought. The surveyor's range finder is shown diagramatically in Fig. 5.7.1; it is held in the left hand in approximately horizontal position and sighted at the target through the eyepiece. Focusing is accomplished by

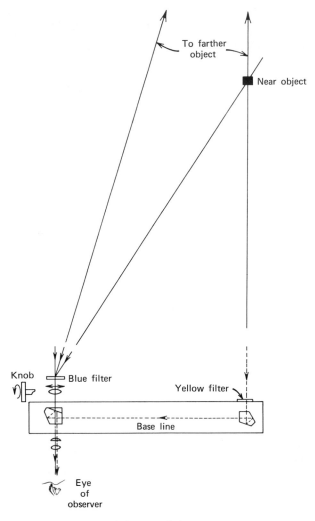

Fig. 5.7.1 Principle of the rangefinder.

turning the knob on the eyepiece. A double image will be seen with one part in yellow and the other in blue. The blue image can be moved laterally by turning the range scale ring while keeping the yellow image approximately in the center of the viewing area. When the two images exactly coincide, the range scale ring will indicate the correct distance— either horizontally or diagonally if the measurement is made along an inclined plane. Its accuracy is somewhat limited because of its short base, but its advantage is that it is a self-contained unit. Such an instrument has very limited use except for the very roughest measurements, typically for reconnaissance.

5.8 Distance Determination by Measuring the Time-Lag of Sound

Infrequently used, a crude but simple method is readily available for roughly determining distance by timing the travel of sound in air. Using walkie-talkies and a stop watch, the sound of an exploding 1-1/2 in. firecracker can signal the starting and stopping of the watch held by a distant observer as he hears it first by radio and then through the air. An average of three or four explosions can give a fairly reliable distance measurement, within perhaps ±10 ft in a mile.

Sound travels at 331.36 m per sec at 0°C at sea level, (1128 ft per sec at 68°F at sea level). Corrections for altitude and humidity could also be introduced, but since the method itself is not extremely exact, the use of temperature corrections only will give answers not too far wrong. See Table 5.8.1 for corrections.

Table 5.8.1   Velocity of Sound in Air at Sea Level

Temperature (°F)	Velocity (ft/sec)	Temperature (°F)	Velocity (ft/sec)
10	1064	60	1119
20	1075	70	1130
30	1086	80	1140
40	1097	90	1151
50	1108	100	1162

Such a procedure might conceivably be used by a field party doing preliminary exploration for sighting transmission-line towers or a highway in a wooded area. The results could prove adequate for such a reconnaissance endeavor. Applications over water may suggest themselves also.

5.9 Distance by Electronic Distance Measuring (EDM)

For distance measurement, the modern surveyman has also new precise devices, any of several new electronic distance-measuring devices (EDM, for short) that may very shortly replace taping. Several makes and models exist, all capable of accurate and precise measurement, which use radio-frequency or light-frequency electromagnetic waves. Those using radio waves need a sending unit and a receiver-transmitter (transponder). Those using light waves require a sending unit at one end and at the other merely a reflector, normally a pentagonal prism. EDM can be much faster and cheaper than taping under certain circumstances, and less liable to mistakes.

5.10 How EDM Works

With any type of EDM equipment, waves are transmitted to a target point and returned, with the time lapse measured and converted to distance through knowledge of their speed of travel. The principle is to compare the unknown distance with the known length of the wave used. The carrier wave can be maintained, typically, by frequency-stable quartz crystals or by some other precise means. The emitted wave pattern is matched to the returning wave pattern and the elapsed time is measured by an electrical delay circuit that can be calibrated with great accuracy. Thus the time of travel to and fro is found, and hence the distance to the target or transponder at the other end of the line.

5.11 Advantages of EDM

Generally, electronic distance-measuring equipment is simpler, faster, and more reliable than taping. None of the devices is too large or cumbersome to be carried into the field. Though costly items, their use will result in an ultimate saving if much measuring is to be done, simply because of their speed and reliability. Any job requiring distance measurements between 100 ft and 1 mi or more can, given a clear sight, be done better by EDM than taping. Traffic, tall brush or grain crops, rough terrain, swamps, bodies of water, or farmland do not interfere, since the sight can be raised above small obstructions and there is no need to set foot on the line except at the ends. Specific applications for EDM are boundary (property) surveying, control surveying, measurements to locate aerial survey points that appear on photographs, measurements at

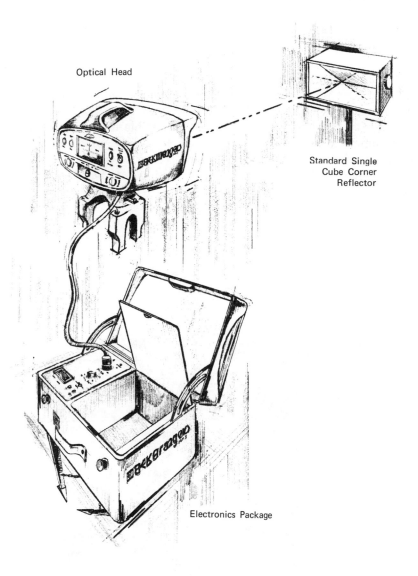

Optical Head

Standard Single
Cube Corner
Reflector

Electronics Package

Fig. 5.12.1 Typical arrangement for electronic distance measurement (EDM) equipment utilizing infrared light mounted on theodolite.

101

bridge or tunnel crossings for precise distance, and so forth. Even instant measurements to moving targets such as hydrographic boats making soundings are no longer a problem.

5.12   Types of EDM Equipment

Several long-range instruments, capable of ranges up to 10 or 20 mi were early developed; the shorter-range instruments, capable of measurements as short as 1 m, came later, mainly because of the new gallium-arsenide (Ga-As) diode. This diode makes possible the much higher frequency signals and the narrow beams, which can give much phenomenal accuracy even to short-distance measure. There is continuing development and increased reliability with each passing day. See Fig. 5.12.1 for an example.

Fig. 5.12.2 Aligning an infra-red EDM device on a retro-reflector.

Courtesy Hewlett-Packard.

Fig. 5.12.3 Microwave distance measuring instrument.   *Courtesy Tellurometer, Inc.*

Typical is one piece of EDM equipment that employs an infrared carrier wave generated by a Ga-As diode, run by energy cells able to make up to 200 measurements on a single charge, that can be carried easily by one person and can measure any distance from 1 to 2000 m in 15 sec. It requires no warm-up time, is quickly mounted on a tripod, gives a digital read-out to five digits, and has a maximum error of ±1 cm at any distance. Figs. 5.12.2 through 5.12.4 show typical EDM devices employing microwave and visible light to measure distance.

5.13 Use of EDM for Construction Layout

There are two difficulties in employing electronic distance-measuring equipment in laying out points for construction—that is, for "stake-out" work. The first is that EDM instruments measure between two fixed points, and thus cannot lay out a predetermined distance directly. The

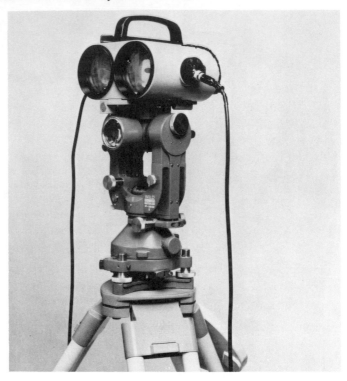

Fig. 5.12.4 Infra-red distance measuring instrument.   *Courtesy Wild Heerbrugg.*

second is that these instruments measure the direct point-to-point distance between instrument and target, thus giving slope distance instead of horizontal distance. Both of these difficulties can be overcome, though, and construction layout ought still to be regarded as entirely feasible with EDM.

To take care of slope correction and reduce any measured distance to horizontal distance requires that the angle of inclination be known between transmitter and target, or that the difference in elevation between the two be known. If the vertical angle be known, then the horizontal distance is found using trigonometry, by:

$$H = S \cdot \cos \alpha$$

where S is the slope distance and α is the angle of inclination.

If the difference of elevation (Δh) is known, then the correction for gradient (as with taping) is found from the calculation:

$$C_g = \frac{(\Delta h)^2}{2S}$$

Fig. 5.13.1 Distance-measuring device with attached vertical circle.

Courtesy Hewlett-Packard.

where S is the measured slope distance and C_g is the amount to be subtracted therefrom to get the true horizontal distance (H) that is desired. Incidentally, since the above formula for C_g is only approximate, it may be well to check the correction by using the fuller version of the formula on steeper slopes to keep from making an error that may be significant:

$$C_g = \frac{(\Delta h)^2}{2S} + \frac{(\Delta h)^4}{8S^3}$$

To take care of the other problem, namely that EDM can only measure a distance but cannot lay it out, the procedure is somewhat simple. One must use a cut-and-try technique. A point is set on line at what is believed to be a good approximation of the distance wanted, and a measurement is taken. Then the point is reset by taping off a correction, and

Fig. 5.13.2 Electronic pocket calculator for conversion of slope distance to horizontal.

Courtesy Hewlett-Packard.

the EDM device makes another measurement. With a few trials, quickly done by the instrument, the desired point can quite readily be laid out. It should be remembered, however, that there is most frequently a slope correction involved, which will have to be made each time the EDM measurement is made.

Despite these two difficulties, however, it is quite practical today to use EDM to lay out construction work, especially where taping would be difficult and time consuming. It often takes much less time and effort to do many EDM repetitions than to perform the taping chore. To facilitate angle measurements at the same time that EDM distances are being done, some EDM devices are attachable to transits or theodolites, so the horizontal angle, the sighting of line, and the vertical angle can be done from the same setup when the distance is being measured or laid out. A correct difference of elevation and a correct horizontal distance are therefore almost immediately available through a simple calculation. Some EDM equipment is manufactured with a small computer attached that converts the slope distances accurately and quickly into true horizontal distances, although these get to be a bit elaborate and expensive. At all events, electronic distance-measuring equipment can finally be used for construction work—if sufficient use can be anticipated for the equipment to prove economical.

5.14 The Laser Beam for Measuring Distance

The laser beam is a modern-day development that can serve as an alignment or, if modulated to a wave form, as a distance-measuring device. It assuredly will become used in many more ways in construction. The laser is an intense beam of highly monochromatic, coherent light. Being coherent and pure, it can be concentrated in a narrow ray and will not scatter in all directions like ordinary light. It can be projected for great distances, can be seen as an illuminated spot on a target placed to intercept it, and can be reflected by a plane mirror or by a prism as desired.

One of the more rapidly developing techniques is that employing the laser, whether the "helium-neon," "gallium-arsenide," or other types. Mention was made of its being used to measure distance; it will be further spoken of as a means of setting and maintaining alignment. At this juncture it may be well to examine this new tool of the surveyman that seems so well suited to surveying and alignment and that is finding new uses every day.

5.15   The Laser Principle

The laser, barely 10 years old, has greatly improved optical measurement techniques for distance, velocity, departure from reference axis, angular rotation rate, and so on. What makes it possible is that laser light is so coherent that it is easily manipulable. What coherence means is that laser oscillations remain in step over long periods of time and that laser output waves remain in step along and across the direction of propagation. The result is that a large percentage of the available energy is concentrated to form an intense output beam of light. The coherence property permits measurements over great distances by optical phase comparison and introduces to the optical region techniques that were previously operable only at radio frequencies. Pulsing, as used in radar and sonar, permit the laser distance to be measured to a reflecting object by measuring the round-trip time of travel. Or, distance measurement can be accomplished with a continuous wave laser transmitter, with a phase-difference measurement translated into distance. (This is the principle of the laser altimeter, very much like the radar-profile recorder.)

5.16   Laser EDM Equipment

By using a helium laser light source in electronic distance measuring, several advantages are gained. The distances that can be measured are virtually from 50 ft to 30 or 40 mi; the light is relatively insensitive to meteorological errors caused by temperature, pressure, and humidity variations along the line, and the power consumption is so moderate that remote or isolated lines can be measured without carrying heavy batteries. The receiving system is provided with a filter that shuts out all light with a wavelength other than that of the signal, so it can be used either in daylight or after dark. A typical system using a helium laser works with a wavelength of 6328 Å, held to frequency stability of better than one part per million by means of a crystal-controlled oscillator. Thus, accuracy and range plus reliability add up to economy sufficient to replace taping in many instances.

5.17   Laser-Equipped Theodolite

A combination of distance measuring and angle measuring is now possible in a single instrument. The Geodimeter Model 700, a long-range EDM instrument, uses a visible helium-neon laser beam good up to 2 mi with a three-prism reflector, and is mounted on a theodolite. Readings of distance and readings of angles (horizontal and/or vertical) can be electronically displayed for read-out. Not too surprisingly, with a built-in computer, the distance displayed can be the calculated horizontal distance. Some manufacturers have managed also to incorporate within the distance-angle computer a device to encode this information on a punched paper tape ready at the end of the task for immediate office processing with no manual transcribing of information. The laser beam can also be simply mounted on its own bracket or tripod, or affixed to a transit or theodolite. In the latter case, the laser may be designed with its own collimated lens system that can be made parallel to the collimation line of the transit, or it can even be arranged atop the transit telescope so its beam can be reflected by a double prism device through the transit telescope and utilize the existing optics. It can be anticipated that this somewhat simpler clamp-on laser will be in vogue for most day-to-day construction alignment operations for some time to come.

6

Laying Out Angles

6.1 Laying Out Right Angles by Use of a Square

In construction work, the 90° angle is frequently used and must be laid out many times. A carpenter's square can be used with sufficiently sure results in many cases on a construction site. As in Fig. 6.1.1, from a base line marked by a mason's line stretched temporarily, the carpenter's square can be used to set a line at right angles.

If occasion demands, a light wooden triangular frame can be made easily from three pieces of lumber to serve for repeated right-angle measurements. (See Fig. 6.1.2.) Occasional checking of the angle and sighting for straightness of the edges is recommended to maintain the frame true.

6.2 Laying Out Right Angles by a Tape

A right angle can be laid out, using a 3-4-5 principle, by a steel tape. The tape is held carefully by three people with the three sides selected to be

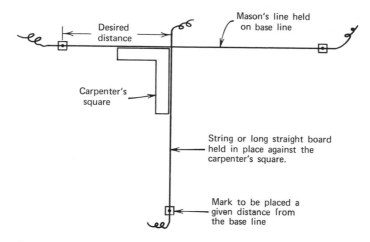

Fig. 6.1.1 Carpenter's square to set a right angle.

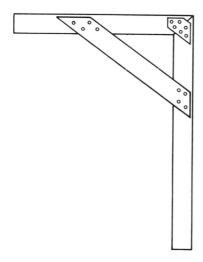

Fig. 6.1.2 Timber frame for laying out right angles on job site.

multiples of 3, 4, and 5 ft, respectively. Figure 6.2.1 suggests adequately how this is done.

Another method of laying out a right angle by using a tape is that suggested from geometry. At point C of Fig. 6.2.2, measure a given distance toward A and mark a point D, and the same distance toward B and mark the point E, say by 10 penny nails pushed into the ground. Then swing arcs of sufficient radius and (using a nail) scribe marks on the

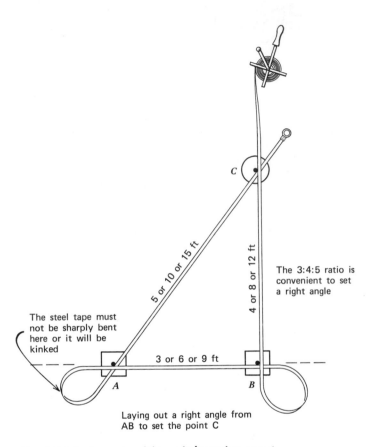

The steel tape must not be sharply bent here or it will be kinked

5 or 10 or 15 ft

4 or 8 or 12 ft

The 3:4:5 ratio is convenient to set a right angle

3 or 6 or 9 ft

Laying out a right angle from AB to set the point C

Fig. 6.2.1 Laying out a right angle by taping, case 1.

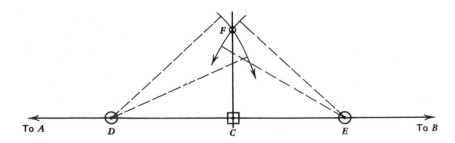

To A

D

C

E

To B

Fig. 6.2.2 Laying out a right angle by taping, case 2.

ground to find point *F*, which is the intersection of the two arcs. Then *CF* is perpendicular to line *AB*, and the lengths of arc used can be of any convenient dimension.

6.3 The Optical Square or Right-Angle Prism

A greatly neglected pocket device for setting a right angle is the optical square. At its simplest, it is a cylinder (Fig. 6.3.1) with slits to form sight lines; more elaborately, it can be a prism that enables one to view the two points *A* and *B* simultaneously and direct an assistant to move right or left until he appears on line. This right-angle prism can be set atop a special pointed stick, called a Jacob's staff, that can be pushed easily into the ground for its temporary task.

The basic principle of an angle prism is that it refracts, reflects, and/ or bends the ray of light (line of sight) at an exact right angle so that an observer can see two points at one time. They appear to be superimposed one over the other. With a minimum of experience, a novice can become rather proficient and inventive, discovering the angle prism to be a frequent checking tool and a new tool for shortcutting.

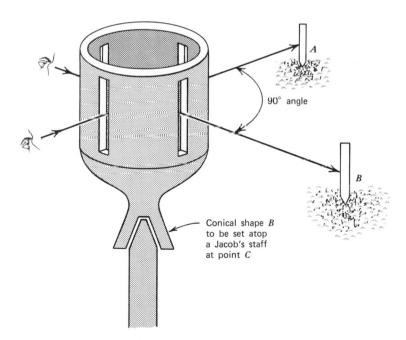

90° angle

A

B

Conical shape *B*
to be set atop
a Jacob's staff
at point *C*

Fig. 6.3.1 Optical square on Jacob's staff to set a right angle.

6.4   Use of the Angle Prism

Angle prisms, virtually pocket surveying instruments, can serve many uses in preliminary surveys, information surveys, and even in construction layout. A single prism can be used to sight in a right angle to one side or the other, as indicated in Fig. 6.4.1.

The prism is hand held, and a plumb bob can be suspended beneath it. Or, for a steadier sighting, it can be held atop a pointed plumb rod that serves as a plummet, a steadying support, or a Jacob's staff pushed into the ground or other independent support.

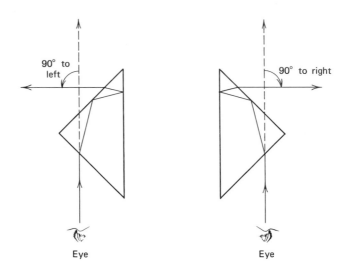

Fig. 6.4.1 The simple angle prism to set a right angle.

6.5   The Double-Angle Prism

A double prism enables one to sight two opposite right angles at once, and is thus an alignment tool; it enables the observer to move himself onto line by watching the two end points of the line until they coincide. Compare Fig. 6.5.1 to the method in Case 3 of Section 4.8.

6.6   Use of the Double Angle Prism

From the sketch it is seen also that one could, once he is on line *AB*, move himself along the line until he sighted himself opposite some de-

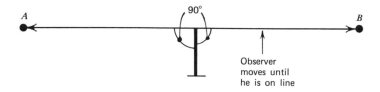

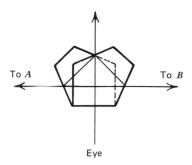

Fig. 6.5.1 Double angle prism for finding a point on a line.

sired point, which he could then mark. He would by this method in effect have dropped a perpendicular to line *AB* at point *C* from any point *D*. (See Fig. 6.6.1.)

He could in this manner stake several points on line *AB* that could be subsequently taped for stationing and the distances to the various objects could be measured from the stakes as desired. All that is needed in this type of operation is, of course, that points *A* and *B* be visible and marked, say, by range poles.

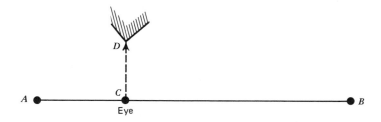

Fig. 6.6.1 Observer at *C* "dropping a perpendicular" from *D* by moving along *AB* with a double prism.

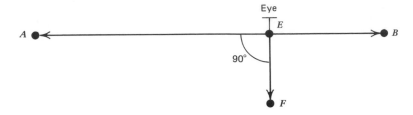

Fig. 6.6.2 Setting a line perpendicular to another line by use of a double prism.

The observer, when setting himself on line *AB* by observing *A* and *B* with his double prism could also sight and set any desired point *F* perpendicular to line *AB* at point *E*, as shown in Fig. 6.6.2.

6.7   Laying Out Angles Without the Transit

To lay out any angle (other than 90°) without using a transit, two methods are suggested.

The first consists of measuring a convenient distance (say 10 ft) along one leg of the angle, erecting a perpendicular there, and then measuring out a distance along the perpendicular ascertained from the tangent of the desired angle. Figure 6.7.1 shows the layout of the angle 28°10′ at point *C*. It requires a table of tangents. The line *CE* is then at the desired angle from line *AB*. For greater accuracy a length greater than 10 ft can be used.

A second method to lay out any desired angle at *C* utilizes trigonom-

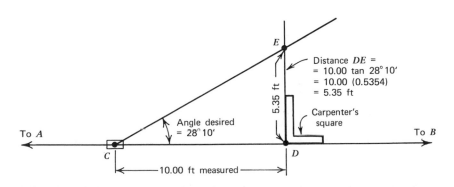

Fig. 6.7.1 Laying out any angle by taping, case 1.

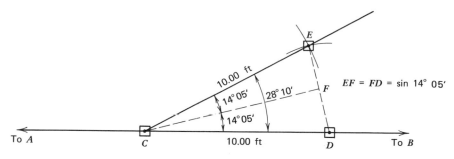

Fig. 6.7.2 Laying out any angle by taping, case 2.

etry also, this time the sine of half the desired angle. From C there is laid off a convenient distance toward B (say 10.00 ft), and an arc is scribed on the ground at the same distance from C, as shown in Fig. 6.7.2. The angle in this example is 28°10′, also.

Since $EF = FD = 10.00$ (sin 14°05′) = 10.00 (0.2447) = 2.447 ft, then the distance to be laid off from D is twice this or 2(2.447) = 4.89 ft. This 4.89-ft arc is laid out from D to intersect the 10-ft arc scribed from C, thus fixing the point E. The line CE is thus at an angle of 28°10′ from line AB.

6.8 Laying Out Angles with the Transit or Theodolite by Repetition

To provide very great accuracy in setting a base line at any angle to some existing reference line, a transit or theodolite is used and a technique is employed that involves repeated measuring of an angle. Essentially it consists of first laying out the angle, then measuring it carefully, finally moving the newly set point as needed to achieve the proper angle. The illustrative example (Fig. 6.8.1) shows a typical procedure. (Compare this with Fig. 6.7.2.)

The problem is to lay out a very accurate angle of 28°10′00″ and set point E 941.00 ft away. The line CE is to serve as the base line for the construction. The transit is used to set E by turning 28°10′ from B, and a tentative mark at E is set on a firm hub or solid surface. Thereupon, a succession of six repetitive angular measurements ("angles by repetition") is made at C, three direct and three reverse, to compile six times the angle, as shown in the notes (Fig. 6.8.2). Division by six establishes that a slight angular discrepancy (β) exists and a lateral shift of point E is required. The amount of the lateral shift is calculated from the size of

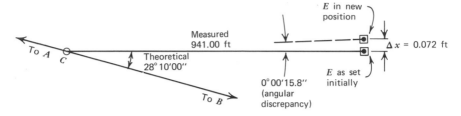

Fig. 6.8.1 Laying out an angle with transit by repetition.

angle β and the length of CE. Knowing that the sine or the tangent of a small angle varies with the angle, and the sin $01' = 0.000291$, the proportionate value of the sin or tan is β (in minutes) $\div$ $01'$ simply. The point E is then shifted laterally by the calculated amount (Δx) and firmly fixed. This calculation is shown in Fig. 6.8.2.

| Point | | Number of Repetitions | | Vernier Readings | | | Angle Value |
Occ.	Obs.	titions	Tel.	A	B	Mean	Accumulated
C	E	0	Dir.	0°00'40''	50''	45''	0°00'00''
	B	1	Dir.	28°11'20''	—	—	(28°10'20'')
	B	3	Dir.	84°31'30''	50''	40''	84°30'55''
	B	6	Rev.	169°02'20''	20''	20''	169°01'35''

Total accumulated value $\div$ 6 = 28°10'15.8''
Lateral correction of point E (Δx) = 941.00 (tan 15.8'') = 941.00 (15.8/60)(0.000291)
= 0.072 ft to be made counterclockwise from C.

Fig. 6.8.2 Form of notes and calculation for layout of an angle by repetition.

Selecting a method of laying out angles depends on the nature of the work being controlled. In general, earthwork or less significant control marks will not require very accurate and time-consuming measurements. Permanent concrete structures and the control grid or center lines on which a large project is based, however, do demand accurate angle layout and measurements by more precise equipment.

7

Stakes, Hubs, and Control Marks for Construction

7.1   Stakes and Hubs

A stake is a piece of wood, typically $1 \times 2 \times 18$ in. sharpened at one end for driving into the ground. The length of the stake varies with the hardness of the ground and with the practice of different people or organizations. A hub is also a stake, usually a 2×2 in. piece of wood, although the length may vary. It needs to be driven firm, to nearly its full length, and a stake (or a lath) can be driven alongside for identification, information, and flagging purposes.

Tall stakes or poles are used where brush or tall grass would obscure a shorter stake. Range poles (surveying poles) painted alternately red and white are for temporary use, accurately fabricated and thus too expensive to be left as permanent sights. A backsight with an instrument is a pointing to a hub or mark set previously, for reference; a foresight is a pointing to a hub or mark being driven or being set in the direction the work is moving.

Fig. 7.1.1 Backsighting on a reference hub.   *Courtesy Kern Instruments.*

Driving a stake requires care to assure that it stays on line during driving, that it is driven to firm bearing, and that it is deep enough that it will not be shifted by passersby or construction traffic. Soft ground may dictate a much longer stake; rocky ground may require that the rock be uncovered and marked with a chisel—or, if a stake is driven, it may have to be guyed, braced, or mounded with rocks.

Where a stake cannot be driven, as on a hard roadway surface, a cross or mark is made with a sharp tool, paint, crayon (keel), or even chalk. Sometimes a nail or tack is driven into the pavement as a mark. Reference indications can be lettered on the curb or walk, on the pavement, on a nearby wall, and so on, giving directions on how to find the mark. These are marked temporarily with lumber crayon or more permanently with paint. Important points should also be recorded in the notes in case the markings are lost or obliterated or the points moved.

Snapping a chalked line onto a floor or a pavement is a helpful means of setting a visible line along which and from which measurements can be made. It can serve the same purpose as the stretched mason's line without creating the hazard of tripping workmen.

7.2   Setting Stakes and Hubs to Control Construction

The hub is driven at correct distance and in correct alignment, then marked for line and distance. Then a guard lath is driven to protect and to "flag" the hub, and serve for recording information for the user of the hub. The placement of construction control points requires hubs to be placed on line and at correct distances from traverse hubs. By taping from the transit hub, distance is fixed; by properly pointing the telescope of the transit, alignment is fixed. Figure 7.2.1 shows the method to be used.

To set a stake or hub on line, the transitman directs the stakeman to move right or left till the point of his stake is on the vertical cross wire, then signals him to drive it. He can watch the stake through the telescope to keep it on line. Then, for placing a point on top of the stake, the transitman signals the stakeman to make a pencil mark at the forward edge and then at the aft edge of the stake. These marks are connected to set alignment on the stake (hub), as seen in Fig. 7.2.2. A stake tack or very small nail can then be driven on line at the center of the stake, but this will be done only as may be determined by taping a correct distance from a previous hub. Identification information is put onto the guard stake.

In the case where driving stakes or hubs is impracticable and nails can perhaps be driven into asphaltic pavement, driving the nail through a bottle cap or a special survey disc will render it more visible; a paint or keel mark can also serve the purpose. Identification information can then be painted on the pavement.

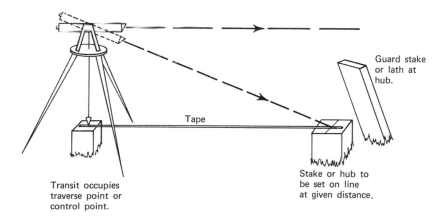

Guard stake or lath at hub.

Tape

Stake or hub to be set on line at given distance.

Transit occupies traverse point or control point.

Fig. 7.2.1 Using transit and tape set a stake.

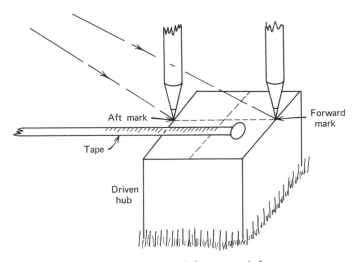

Fig. 7.2.2 Detail of marking line and distance on hub.

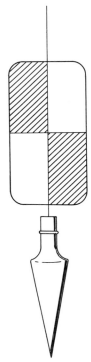

Fig. 7.2.3 Plumb bob with string target for sighting from transit.

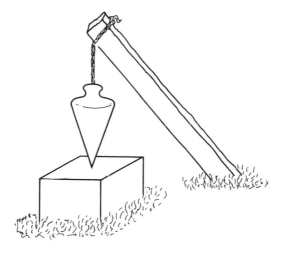

Fig. 7.2.4 An improvised backsight for the transit man.

Sighting on a hub often requires an assistant to hold a nail or pencil or plumb bob over the mark so the transitman can orient his instrument. A plumb bob cord held over a hub at a great distance can be made more visible to the transitman if a plumb bob string target is used (See Fig. 7.2.3) A resourceful assistant can perhaps drive a stake or lath, slanted over a hub, in such a manner (Fig. 7.2.4) that he can have a plumb bob hang over the point for frequent sighting by the transitman. This will release him for other duties.

7.3 Auxiliary Sighting Targets to Fix Line

If frequent backsighting over a long period of time will be needed during layout, a permanent device can be set over the particular hub to be sighted, or a target farther along the line can be affixed to a wall, on a taller stake, or on a building for convenience. Figure 7.3.1 shows some examples. Painting a target or using lumber crayon (keel) and refining it with a sharp pencil can adequately serve the purpose. This will eliminate the time lost in sending a person out each time to hold a pencil or plumb bob on a hub, and also cut down on the ever-present risk of his carelessly using a wrong hub.

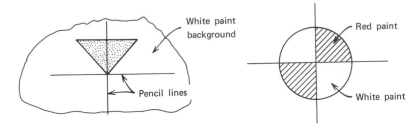

Fig. 7.3.1 Typical temporary sighting targets painted on wall.

7.4   Permanent versus Temporary Control Points

Monuments and bench marks are permanent markers, usually concreted in place. They can be simple brass or steel rods or discs anchored so as to protrude slightly above ground for reference subsequently. They mark alignment of property lines or control surveys, or give fixed elevations.

A prudent precaution is to carry the vertical and horizontal control on both permanent and temporary points. The permanent marks, set back from the active construction area, are then frequently checked with the temporary "working" control to detect immediately any deviations. It will save time, expense, and embarrassment that might occur from error resulting from disturbance to control points exposed in the work area.

7.5   Setting a Permanent Hub from a Temporary Stake

If a hub sometimes needs to be replaced by a more permanent monument, or if it needs to be removed temporarily in the course of construction, it can be set exactly in the correct original position if reference marks are initially placed. The simplest means is to set two pair of marks forming two lines, with maybe a third pair giving a third line as a safety measure. To reset the disturbed hub one stretches mason's line across the pairs of reference marks, and the point where the lines cross is the point sought. The third line will serve as a check; it may turn out to be essential if one of the original four reference marks is moved or destroyed by accident during the operation. Figure 7.5.1 shows the process. If solid reference stakes are driven initially, a nail can be driven into the top of each one so the strings are properly aligned over the hub. Later the mason's line can be tied to these nails and a plumb line can be used to set the point on the hub. This makes it convenient for one person to

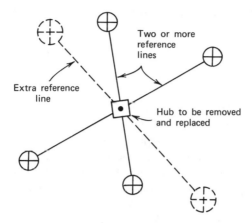

Fig. 7.5.1 Removing a temporary hub and replacing it by use of reference lines.

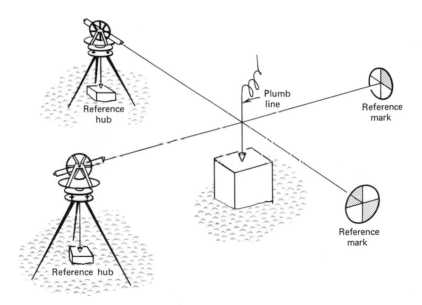

Fig. 7.5.2 Hub being replaced in its original plan location by use of transit(s).

do the whole job. Sometimes it is convenient to reestablish a disturbed hub by two lines of sight from two transits. Figure 7.5.2 depicts this procedure.

7.6 Types of Permanent Control Marks

Permanent markers to replace stakes may be required to assure that the survey will not be lost or destroyed during a prolonged planning stage. For these more durable control points, monuments can be set into concrete or stone. Some examples are shown in Figs. 7.6.1, 7.6.2, and 7.6.3. Other types exist, including some with buried steel shafts that can be found with a magnetic dip needle or other type of magnetic probe.

The "permanent" marks one sets for later use are to be used with caution when later recovered. Whenever a bench mark (or hub) is reused after some time lapse, it must not be assumed correct until checked. This is done by running levels from the bench mark over one or more other known elevation points, or measuring to align the hub from one or more fixed points for verification.

For example, nails driven into asphaltic concrete can creep with the pavement under the long-term action of traffic. Crosses or marks on flagstones can move if the flagstone is lifted and reset or run over by a

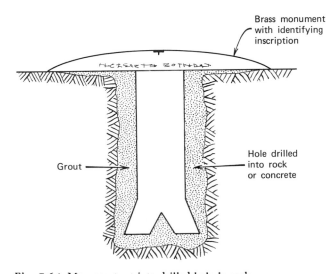

Fig. 7.6.1 Monument set into drilled hole in rock.

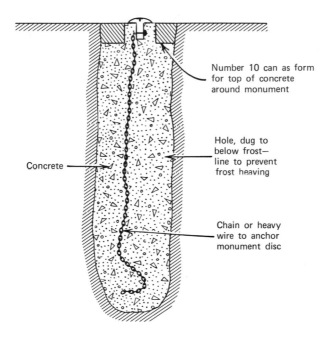

Number 10 can as form
for top of concrete
around monument

Hole, dug to
below frost—
line to prevent
frost heaving

Concrete

Chain or heavy
wire to anchor
monument disc

Fig. 7.6.2 Monument set in ground.

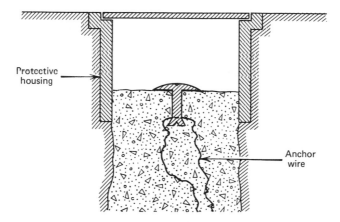

Protective
housing

Anchor
wire

Fig. 7.6.3 Monument in street set in a protective housing below pavement.

construction vehicle. Frost heave during the winter can disturb a mark; it can change the elevation of a bench mark appreciably. If manhole rims are used for permanent reference, there can also be problems if the manhole rim is raised and reset to accommodate a pavement resurfacing. It even happens that overly helpful people remove bench marks or property monuments "to prevent their being damaged" by construction activities, and later replace them, unwittingly causing blunders for subsequent users.

8

Coordinate Surveying and Layout

8.1   Coordinates and Grids

Construction is very adaptable to being set out by coordinates on a grid system. The coordinates are the x and y locations from a set of axes, as in algebra: the x-coordinate is the distance from the y-axis, and the y-coordinate is the distance from the x-axis. Mostly the y-coordinate is called the N-coordinate, and the x-coordinate the E-coordinate. If the construction is located in the northeast quadrant, as it mostly is, then both the x and the y are positive or plus. If the N-coordinate becomes negative, it becomes an S-coordinate; if the E-coordinate becomes negative, it becomes W-coordinate. See Fig. 8.1.1 as an example.

When the N-coordinate and the E-coordinate are given for a point, the point is fixed with respect to a rectangular coordinate system, and a rectangular grid can theoretically be established on which any such point can be located or placed. It is obvious that a plane rectangular grid will properly fit the earth's shape (the oblate spheroid, or geoid) only over a relatively small portion, but since the usual construction project covers only a small area, the rectangular coordinate grid works well.

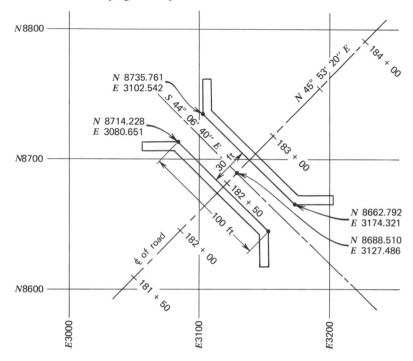

Fig. 8.1.1 Highway overpass structure with coordinates shown for key points (scale: 1 in. = 50 ft).

To find a particular point on a map or plan from its known coordinates is a simple matter of scaling. One need merely measure off the distances on the plan from the *x*- and *y*-axes from the nearest rectangular grid lines. So, too, one can discover the *N* and *S* coordinates of a point by scaling the map or plan—although these coordinates will be only as accurate as the scale of the plan will allow. (See Fig. 8.1.2.)

To set out a point on the ground, however, by knowing only its *N* and *S* coordinates is quite another matter. But it is this very procedure that is regularly done; construction points are placed mainly from their calculated coordinates. It requires that the surveyman perform the necessary computations and then measure off his computed distances and directions to the ground points that are wanted. The actual *x*-axis and *y*-axis for a project are not visible on the ground, nor are any of the grid lines. The best that can usually be had to work from are some control points whose grid coordinates (*x* and *y*, or *N* and *E*) are known. One must work from these known points to establish the new points that locate the construction project.

8.2   The Surveying Problem of Finding Coordinates of Points

To understand the problem we must recognize relationships and terminology used in traverse surveying. In Fig. 8.2.1, the Δy of AB is the "northing" or "latitude" of AB, the north-south distance that the course AB extends; the Δx of AB is its "easting" or "departure," the east-west distance it extends. To calculate the northing (Δy) of this course, one multiplies the course length by the cosine of the bearing; to calculate the easting (Δx), one multiplies the course length by the sine of the bearing.

For example, line AB has a bearing of N 81° 15' 10" E and a length of 200.00 ft. Thus

$$\Delta y = 200.00 \text{ (cos Brg.)} \qquad \Delta x = 200.00 \text{ (sin Brg.)}$$
$$= 200.00 \text{ (0.15207)} \qquad = 200.00 \text{ (0.98837)}$$
$$= 30.41 \text{ ft northerly} \qquad = 197.67 \text{ ft easterly}$$

If the coordinates of A are given, for example, as N 3121.74 and E 2874.76, then we now find the coordinates of B:

N 3121.74	A	E 2874.76
+ 30.41	AB	+ 197.67
N 3152.15	B	E 3072.43

For a succession of courses as shown in Fig. 8.2.2, the total calculation can most conveniently be done in a tabulation, using (in Table 8.2.1) natural sines and cosines and a desk calculator. Coordinates are used extensively to fix construction control points exactly in a grid system.

Table 8.2.1   Coordinate Calculation

Points Lines	Bearings Lengths	Cosines Northings	Sines Eastings	N-Coordinates	E-Coordinates
A	—	—	—	N3121.74	E2874.76
AB	N81-15-10E	0.15207	0.98837	—	—
	200.00	+ 30.41	+ 197.67		
B	—	—	—	N3152.15	E3072.43
BC	N25-34-15E	0.90205	0.43162	—	—
	150.00	+ 135.31	+ 64.74		
C	—	—	—	N3287.46	E3137.17
CD	S47-23-30E	0.67798	0.73600	—	—
	175.00	− 118.65	+ 128.80		
D	—	—	—	N3168.81	E3265.97

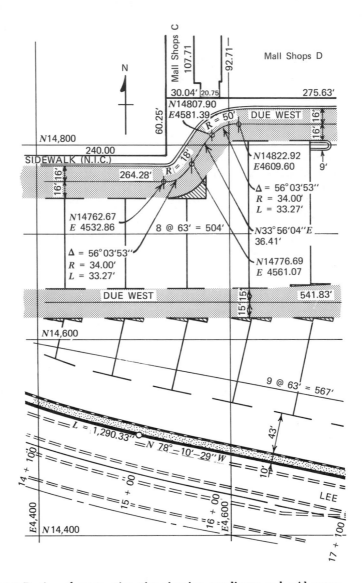

Fig. 8.1.2 Portion of construction plan showing coordinates and grid system.

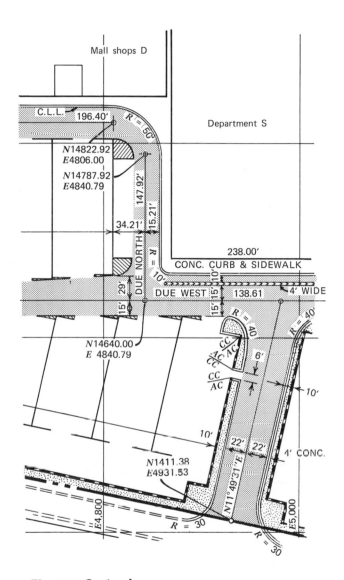

Fig. 8.1.2 Continued.

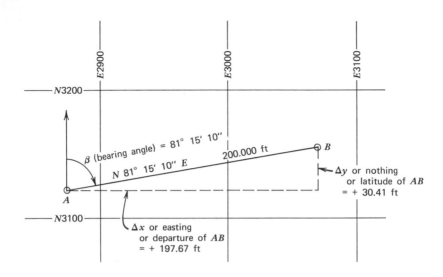

Fig. 8.2.1 Latitudes and departures parallel to grid axes

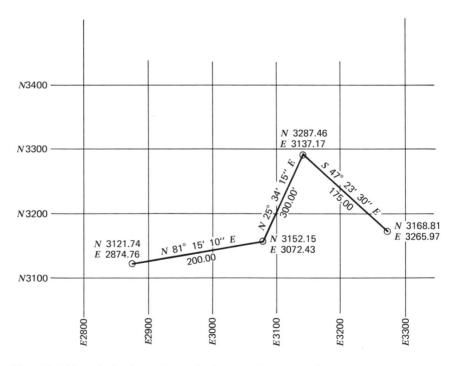

Fig. 8.2.2 The calculated coordinates for the control traverse points.

134

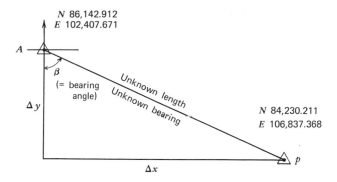

Fig. 8.3.1 The inverse solution.

8.3 *The Inverse Problem of Using Coordinates to Find Points*

The foregoing is a commonly encountered computation to find the coordinates of any significant points for plotting them on a map or for finding them in the field. There is an opposite computation, the "inverse" operation, which is very important for laying out points in the field when their coordinates are known. One must then work from a known control point to set out an as yet undiscovered point by knowing only coordinates, not distance and bearings. The purpose of the inverse problem is to find the distance and bearing to the point whose coordinates we are given, as in Fig. 8.3.1, so the point can be set in the field.

For example, to compute the bearing and distance to a point (P) of known coordinates, one works from another point (A) whose coordinates are known, finding first the latitude (Δy) and departure (Δx) to the new point. Then one divides Δx by Δy to find the tangent of the bearing angle (β).

$$\tan \beta = \frac{\Delta x}{\Delta y}$$

where Δx = difference in *E-W* coordinates ("eastings") and Δy = difference in *N-S* coordinates ("northings"). In a table of natural trigonometric functions, the value of angle β is found (to degrees, minutes, and seconds). At once the sin β and the cos β are found from the tables (using interpolation) for use in calculating the distance:

$$\text{Distance } AP = \frac{\Delta x}{\sin \beta} = \frac{\Delta y}{\cos \beta}$$

Both divisions are performed, for confidence.

As an illustrative example, an inverse computation is done here by several methods. Given the coordinates of A and of P, it is desired to find the distance and bearing of the line AP. Here is the calculation done to eight digits, using natural trigonometric function tables to eight decimals, and worked on the desk calculator:

N-Coordinates		E-Coordinates	
A	N86,142.912	A	E102,407.671
P	N84,230.211	P	E106,837.368
$\Delta y =$	$-1,912.701$	$\Delta x =$	$+4,429.697$

$$\tan \beta = \frac{\Delta y}{\Delta x} = \frac{+4,429.697}{-1,912.701} = -2.31593804$$

(The negative sign gives the clue to the proper quadrant.) From the natural trigonometric function table, find the bearing angle,

$$\beta = 66°38'44.8''$$

Then, having the value of β, find in the tables these values for

$$\sin \beta = 0.91807161 \qquad \cos \beta = 0.39641458$$

Then calculate the length of AP two ways, thus:

$$\text{Distance} = \frac{4,429.697}{0.91807161} \qquad \text{Distance} = \frac{1,912.701}{0.39641458}$$

$$= 4,825.002 \qquad\qquad = 4,825.002$$

Thus the bearing of AP is S 66°38'44.8"E, or its azimuth is Az 113°21'15.2", and its length is 4,825.002 ft.

These two distance values will check each other. If they do not come very close, it is an indication that the interpolation in the tables was not precise enough, perhaps an indication of a blunder in computing. There are not many books of tables of natural trigonometric functions that have more than five digits. A set by Peters, to eight decimal places for each second of arc, does exist and is very useful here. Since the number of digits in Δx and Δy is seven, one ought to use trigonometric tables with that many digits, at least.

Working this problem by six-place logarithms is possible with nearly the same accuracy, though the computation is a bit more tedious. This is the example problem worked with six-place logarithmic tables.

$$\Delta x = +4,429.697$$
$$\Delta y = -1,912.701$$

Log Δx = 3.646374
Log Δy = 3.281647
Difference = 0.364727 (= Log tan β)
Bearing angle (β) is, therefore, 66°38′44.9″

Log Δx	= 3.646374	Log Δy	= 3.281647
− Log sinβ	= 9.962876	− Log cos β	= 9.598149
Difference	= 3.683498	Difference	= 3.683498

Distance is, therefore, 4825.011; Bearing is *S*66°38′44.9″*E*

It will be noted that this result is somewhat in disagreement with the result calculated by the eight-place natural tables, which may very well be expected. Six-place logarithmic tables will conceivably show a discrepancy in the sixth digit of the result; so, too, would six-digit natural function tables or any computation worked only to six digits.

For comparison, this is the example problem worked with seven-place logarithmic tables:

Log Δx = 3.6463740
Log Δy = 3.2816471
Difference = 0.3647269 (= Log tan β)
Bearing angle (β) is, therefore, 66°38′44.78″

Log Δx	= 3.6463740	Log Δy	= 3.2816471
− Log sin β	= 9.9628765	− Log cos β	= 9.5981496
Difference	= 3.6834975	Difference	= 3.6834975

Distance is, therefore, 4825.002; Bearing is *S*68°38′44.8″*E*

Thus far in the illustrative example of this inverse problem, the most reliable result ought to be that using Peters' eight-place tables. The others could be counted upon to be fairly reliable in the sixth and seventh digits, respectively. There is every reason, however, as a practical matter for finding and laying out the point in the field, to round off the result to 4825.00 and *S*68°38′45″*E*.

The same problem done on an electronic computer would not necessarily have a better or more reliable result; it depends on the number of integers used by the machine in its computing of sine, cosine, arctangent, and other operations. This can only be learned by inquiry in each case. If the machine's subroutine calculates trigonometric functions to only seven places, the result of the calculation will be good only to about the seventh digit. This is about the equivalent of handling bearings to the nearest second (or maybe the tenth of second). Roughly, the number of digits in a bearing angle corresponds with the number of digits of capability of the computer's sin, cos, and arctan operations.

A randomly selected bearing of N80°14′37.6″E contains seven digits, which will be seen if one converts it to the decimal form of degrees (80.24377°) or of minutes (4814.627′) or of seconds (288,877.6″). These are seen to be all seven-digit numbers, so in such a case, the computer or the natural-function table or the logarithmic trigonometric table should have a seven-digit capacity. If a six-digit bearing were involved, such as N80°14′38″E, then any computation could be handled with six-digit trigonometric tables, six-place logarithms, and so on.

8.4   Use of the Inverse Problem Solution

Many uses of this inverse problem can occur in construction layout, since virtually all large construction (e.g., a shopping center) is based on a coordinate grid system. It is necessary to set hubs at various points inaccessible one to the other, virtually unrelated one to the other, and known only by their calculated coordinates. To be able, therefore, to discover a distance and direction to any new hub from the known control points (e.g., from traverse points) is very needful. When available, an electronic computer can solve the problem rapidly, but one must be able to do it with natural or logarithmic functions, since the case occurs frequently.

For example, in Fig. 8.4.1, showing a segment of a connecting road to be constructed because of a shopping center disruption, the key points are shown by their coordinates only. These points must be set in position before the work can begin, and the inverse solution will surely be called for.

8.5   Surveying by Simple Triangulation

To extend horizontal control to a point or points relatively inaccessible by traverse methods, triangulation may be used. In simplest form, a new point can be located, visible from two known control points, and its coordinates can be calculated by merely making angle observations. In Fig. 8.5.1, for example, C can be found from A and B. AB will be the base line, already known from existing coordinates, and the work principally requires measurement of angles about A, B, and C. The angles are measured by repetition with a transit, or by a theodolite, and angular adjustments are made for closure. The distance CB is found by the law

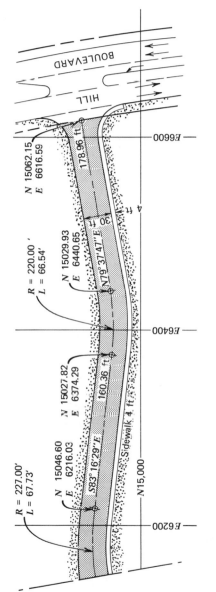

Fig. 8.4.1 Construction points to be laid out from their known coordinates.

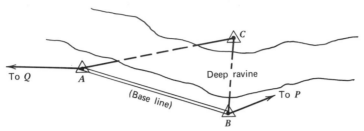

Fig. 8.5.1 Simple triangulation.

of sines; its bearing is worked from the angles; and the coordinates of *C* are found by computing latitudes (Δy) and departures (Δx) for *AB* and *BC*, as follows:

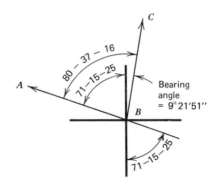

Fig. 8.5.2

AB = 767.810 ft (known)   Bearing of AB: $S71°15'25''E$ (known)
Known coordinates of A: N 817,621.87 and E 248,412.32
Angles from measurement: Adjusted angles:

A =   17° 16' 08''   17° 16' 12''   (The 12'' discrepancy
B =   80 37 12   80 37 16   is divided equally
C =   82 06 28   82 06 32   in this example.)

 Sum = 179° 59' 48''   Sum = 180° 00' 00''

Discrepancy = 12''

From the law of sines,

$$BC = \frac{\sin A}{\sin C} \times AB = \frac{0.29687492}{0.99053077} \times 767.810$$

$$= 0.2997130 \times 767.810 = 230.123 \text{ ft}$$

Bearing of *BC* is figured to be *N*9°21′51″*E*

A					N817,621.87	E248,412.23
AB	767.81	S71-15-25E	0.3213247	0.9469691	−246.72	+727.09
B					N817,375.15	E249,139.41
BC	230.12	N09-21-51E	0.9866741	0.1627089	+227.06	+37.44
C					N817,602.21	E249,176.85

Finally the point *C* whose coordinates have been ascertained by triangulation can now be used like any other coordinated point in the network of control points.

8.6 Setting Out a Point by Intersection

Another insight from this triangulation discussion is that in construction a method exists for setting out a coordinated point by intersection when it may be impractical to set it out by taping. By setting up two transits or theodolites on known points *A* and *B* and orienting the instruments in the proper direction, a new point *D* whose coordinates are known can be set at the intersection of the two lines of sight from the two instruments at *A* and at *B*. (See Fig. 8.6.1.)

The direction of *AD* is calculated from knowing the coordinates of *A* and of *D*, by use of the inverse computation previously illustrated. The direction of *BD* can be similarly obtained from the coordinates. Then the proper angles at *A* and *B* can be used to turn the intersecting lines *AD*

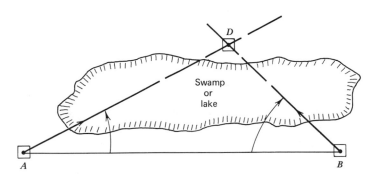

Fig. 8.6.1 Setting a point by intersection of two lines of sight.

and *BD* to set *D* at its proper location. This is a method that lends itself to setting a point requiring over-water measurements, for example, setting a bridge pier, a marker buoy, or a point on an opposite shore.

An occasional need may occur to extend the triangulation through several triangles, as in Fig. 8.6.2, as the most feasible way to set points for construction. The theodolite must occupy each station and measure the angles indicated. Then an adjustment must be made by some means to

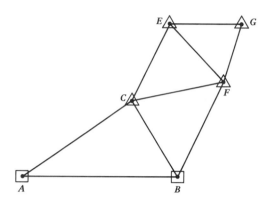

Fig. 8.6.2 Extending control points by triangulation through several triangles.

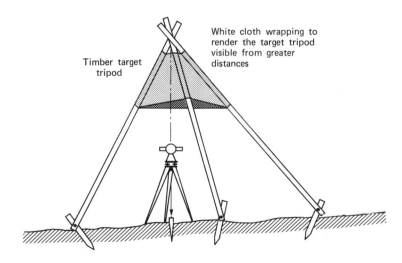

Timber target tripod

White cloth wrapping to render the target tripod visible from greater distances

Fig. 8.6.3 Elevated triangulation target to be observed from a distance (one variation among many).

assure that the sum of all angles about a point total 360° exactly, while at the same time the sum of the angles in each triangle total 180° exactly. Then a law-of-sines calculation must be made, as in Section 8.5, to find the coordinates of the final desired hubs (E and G in Fig. 8.6.2). The need for this type of work in construction layout is rare, but may sometimes occur in difficult terrain with ravines, swamps, or other obstacles that interfere with straight measurement. Targets can be set for the longer sights involved, by building timber tripods tall enough so that an instrument can be set up on the point beneath, an example of which is shown in Fig. 8.6.3, although placing a target on a theodolite tripod and clearing a bit of brush is preferable.

9

Principles of Construction Layout

9.1 Construction Point Information

The location of a point is "where it is"; the elevation of a point is "how high above sea level it is"; these add up to the three coordinates, x and y (location), and z (elevation). These bits of information may be furnished for a point, and the job of the constructor's surveyman is to set the point correctly so work can begin. He will frequently have no more information than the three coordinates for a point, and it is his function to translate these coordinates and elevation into a point properly staked on the ground.

The task for the construction survey crew will normally be to set alignment hubs from existing survey marks, stakes, street lines or curbs, buildings, and so forth, so that construction can be properly undertaken with reference thereto. A center line or base line should be established on the construction site, and elevations are set on typical control stakes (bench marks) needed at once to begin. All control points must be well marked and safeguarded with flags and stakes to prevent construction equipment from bumping them or running over them. Most important, a record must be maintained in a notebook to assure that the job has been correctly done and so that the work can be verified later or retraced and reset.

9.2   Reference Hubs

Another usual practice in construction work is to see that any hubs that will be disturbed are referenced from offset lines or stakes. If the offsets are kept less than 6 ft, a carpenter's rule (and a plumb line) will serve to reestablish the point. Figure 9.2.1 illustrates this simplest method.

If reference lines must be set further back, taping will be required, and frequently also a transit, to reestablish construction hubs. Several methods may be suggested in Section 4.8, 4.9, 6.2, 7.5, and 8.6 for resetting a disturbed hub from reference hubs. Inasmuch as any construction stake will be disturbed at least once, some foresight and ingenuity must be employed to keep replacement work to a minimum and to make it easy to do. Putting "flags" or guard stakes around them will help, both for main hubs and for reference hubs. To use the principle that three points must be used to fix every line will also help. Periodic alignment and position checks ought regularly to be made, too, to avoid the risk that any construction will be carried forward in error and have to be torn out at sometimes considerable expense.

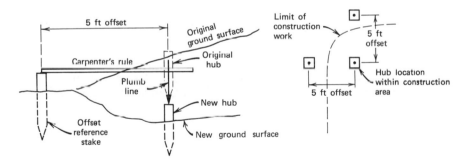

Fig. 9.2.1 Reference stakes for resetting a disturbed control hub.

9.3   Procedure for Laying Out Construction

9.3.1   Horizontal Control

Virtually everything constructed is set out with reference to street lines, pierhead or bulkhead lines, property lines, or base lines of some sort. The layout requirements are fixed by the design engineer or architect, sometimes quite critically. The layout procedure demands that the reference line(s) or base line(s) first be discovered or set beyond question of doubt before any layout measurements can begin. The services of the knowledgeable land surveyor in the area must frequently be enlisted to set boundary lines authoritatively; sometimes the ties will be made to existing structures within a site, making it a different problem.

Most frequently it is advantageous, even necessary, to set up a base line for the construction. A main base line along the central axis of the structure is frequently set, as a succession of firmly established monuments that will (hopefully) endure and serve as references throughout the construction period. For extra security, firm marks are also placed on line at each end, well beyond the construction activity. These are marks that can be occupied and/or can be sighted for verification in the event that marks within the construction are disturbed. The time and effort expended on initially setting the base-line monuments and safeguarding them thereafter will pay off manyfold throughout the job.

Frequently, because construction will require disturbing a central base line, a secondary base line is set parallel to but at some distance from the central axis. It could be offset along the face of the finished building, or one line could be set along the front and another along the rear. Sometimes two equally important axes govern the construction, as in the case of a highway overpass or underpass.

9.3.2   Elevation Control

Elevations are required as well, and vertical control must be established at the construction site so that measurements can be made frequently throughout the job. Many firm, fixed points (bench marks) are labeled with their correct elevations, and others are set and labeled in such areas and in such ways that they will probably not be disturbed during the work. As with the horizontal control, these vertical control marks are supplemented by others set outside the limits of the construction area for security.

For assurance, a periodic level circuit is made of all the bench marks to discover any that may have settled or been disturbed by frost, slides, or heavy equipment. The main or primary bench marks, outside the work area, are used to check the temporary bench marks within the work area. These close-in marks must be used with caution each time. One might always expect they could have been disturbed, and so work from two or three as a precautionary practice when giving levels for parts of the work.

9.4 Typical Survey Points to Control Construction

Control points governing construction operations and survey points are typically:

1. Monuments, points, or hubs marking the elevations of adjacent property.

2. Monuments, points, or stakes marking the corners or elevation of a structure.

3. Offset stakes, points, or monuments.

4. A line or lines through a project, or a base line immediately adjacent to a structure, giving line and grade necessary for the contemplated construction operation.

5. Grade and alignment stakes for utilities, and also other special lines and grades that may be specified.

6. Slope stakes for marking edge of excavation, and alignment marks for piles, piers, and caissons.

7. Vertical and horizontal control points on the various floor levels of a multistory building.

8. Lines and grades necessary for the correlation and location of two or more adjacent structures.

9. Base lines to control highway work.

In general, the carpenters and other trades can then work from such survey lines and marks for all ordinary construction. Experience is the best teacher, and although every job is different, one can quickly learn what marks to set, where best to set them, and how to safeguard them.

9.5 Batter Boards for Construction

Simple construction can be handled by placing better boards to give both horizontal and vertical control (line and grade) at the building corners, along culvert faces, over ditches for placing drainage pipe, and so on. These are simply firmly driven temporary timber frames between which mason's line can be stretched for alignment of the foundation forms, the bricklaying, the pipelaying, and so forth. Chapter 10 illustrates the use of batter boards. Their main purpose is to enable the workmen to readily measure from a reference without need for a surveyman on a standby basis. Subsequent chapters speak of batterboards for use in layout of buildings, sewer lines, and other structures. Figure 9.5.1 is a simple illustration of batterboards to guide construction of a small dwelling.

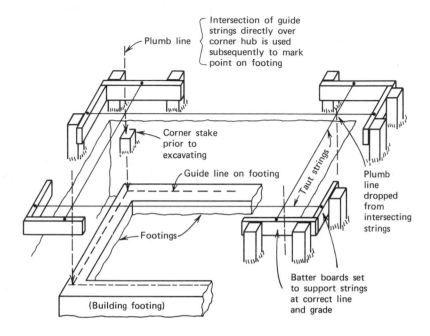

Intersection of guide strings directly over corner hub is used subsequently to mark point on footing

Plumb line

Corner stake prior to excavating

Guide line on footing

Taut strings

Plumb line dropped from intersecting strings

Footings

Batter boards set to support strings at correct line and grade

(Building footing)

Fig. 9.5.1 General schematic to show use of batter boards.

10

Building Construction Layout

10.1   The Building Plot Plan

Before any design can begin, all the available information about the construction area must be discovered and made available to the design engineer-architect team. This is best done on a drawing, made to scale as requested and dimensioned as needed. For building construction, this drawing is called the building plot plan, and sometimes the architect's plan.

To acquire the information for drawing the plot-plan, the surveyman must make measurements for spotting manholes, finding sewer invert elevations, locating utilities, acquiring topographic information, trees, water courses, and so on. Property-line monuments and distances to existing buildings, walls, easements, encroachments, trees, rock outcrops, walks, curbs, and pavement will all be shown. All property-line work is the province of the land surveyor and of the surveyman in his employ. The

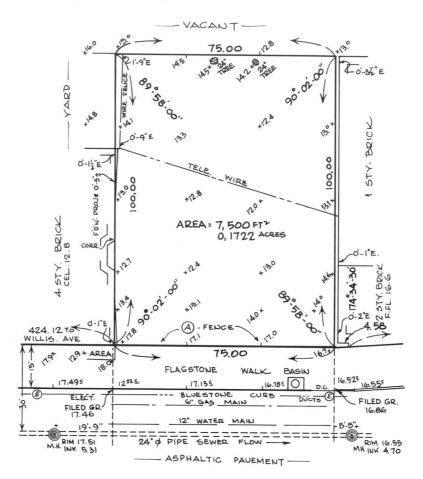

Fig. 10.1.1 Building plot-plan.

land surveyor is the one who certifies the correctness of the property line and related information. (See Fig. 10.1.1.)

Subsequently the engineering-architecture design is based on this plot-plan information, and the outline of the proposed construction is super-imposed on the plot-plan sheet. It is from this layout sheet that the actual layout work is done. Thus the plot plan delivered to the layout survey-man shows the relationship of the new building to property lines (and monuments), street lines, curbs, utility lines above and below ground, and other features. It usually will also show the new contour lines and needed elevation information, such as special bench marks to be established on the site for controlling the construction.

10.2 Setting Corners of Building

In simple building layout, key corner marks are placed on pegs or hubs, using the transit and tape to assure right angles and correct distances. The initial layout on the ground should be assuredly established in correct relationship to base lines, property lines, buildings, and so on. Bringing these control lines to the building site requires care. Usually at this stage the building-site survey will have established horizontal and vertical control for the site. The transit is indispensable for correct angular layout, and even the repetition method should be used for strong construction axes to be truly perpendicular. Steel tapes must be used to assure distance fixes.

As indicated in Fig. 10.2.1, once the building is laid out, diagonals should be measured and compared as a further check that all the building angles are square. For example, *AC* should equal *BD*, and *CF* should equal *EG*.

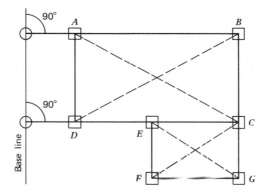

Fig. 10.2.1 Layout of corners of a simple building.

10.3 Batter Boards for Buildings

When setting the corner hubs for a simple building, the corner points should be secured by reference marks set well outside the work area, which can be used to reset the corner marks at needed times. The batter-board technique, however, can furnish a frame for string-lining of the construction sufficiently precise for the trades. See Fig. 10.3.1 for the location of batter boards to control a simple building.

As a general practice, batter boards—wooden frameworks for marks from which to stretch mason's line or wire—are placed to mark the cor-

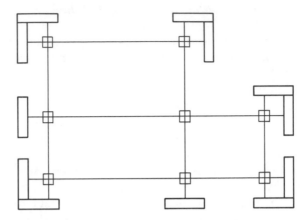

Fig. 10.3.1 Batter boards set to control a simple building, with corner hubs set.

ners. See Fig. 10.3.2 for details. Inasmuch as the corner hubs will be disturbed by the construction (e.g., by the foundation excavation), batter boards have to be fairly permanent so that mason's cord or wires can be stretched to reestablish line and grade whenever needed by those setting concrete forms, steelwork, masonry, and so forth. They must be set back just far enough so as not to be disturbed, yet handy enough so that line

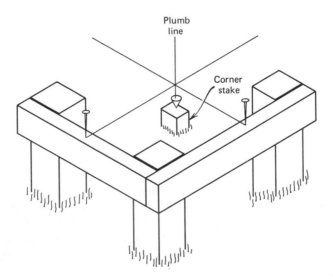

Fig. 10.3.2 Close-up of batter board with reference to a building corner hub.

can be stretched between them conveniently. A plumb bob held at the intersection of two batter strings will reset the corner of the structure (or the center of the piers) when required.

To prearrange that batter boards for a building shall be set at the correct height, usually a convenient number of feet exactly above finish floor grade or above foundation grade, one needs timbers of sufficient length. First three firm timbers are driven at each building corner, about 4 to 6 ft back from the intended excavation. When the engineer's level has brought in the elevation, then the cross boards can be set between them while a level rod is resting on them, to assure their proper elevation. Only then can the nails be set in proper position to hold the strings for corner alignment. This is done by stretching strings over the corner hubs (using plumb bobs as required) or else a transit is used to set the marks on the cross boards. Nails are then driven vertically at the proper location on each cross board.

Since batter boards must give both alignment and elevation, care is needed to set them. As a final assurance, all batter boards must be checked for correctness after placing, sighted from the transit set up successively on the reference points previously set, and verified by leveling once more on the crossboard.

10.4   Construction Base Lines

Buildings, shopping centers, industrial parks, bridges, and other structures are most conveniently laid out from a main base line and an auxiliary base line at right angles to it. Along each base line hubs are set where required for setting or aligning building corners, centerlines, or other significant elements. An illustration of this in Fig. 10.4.1 shows the basic method. In a situation where frequent and repeated alignment is called for on piers, footings, or columns, such alignment ought to be given by sighting directly between points or by stretching wires or strings. It is undesirable to require the setting or turning of angles each time a center point is wanted.

Thus, when base lines are set and hubs marked along the base lines for whatever setting-out may be anticipated, additional auxiliary hubs must be set for future sighting whenever possible. Figure 10.4.2 shows such sighting marks (or hubs) intended to ease the never-ending alignment task. Sighting targets are best affixed to or painted on walls, being less liable to destruction there. The work is set out with hubs and tacks first and the targets are put in after sighting the hairline on the hub tacks.

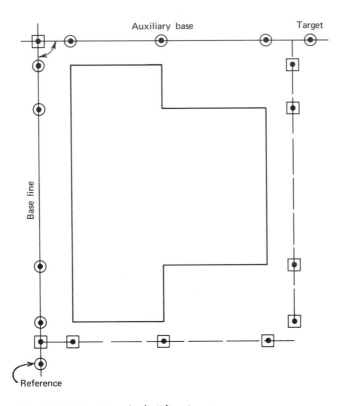

Fig. 10.4.1 Base lines for building layout.

Setting out a base line for columns for a building or structure is illustrated here. The main 90° angle is turned carefully first, as explained in Section 10.5. Survey stakes needed for the building are basically as indicated in Fig. 10.4.2 so that further measurements can then be done by the carpenters and other trades.

10.5 Setting Base Lines at True Right Angles

In the construction layout of Fig. 10.4.2, the two base lines must be set assuredly and beyond question at the correct angle to each other, 90° in this case. Therefore, great care is demanded in setting the angle QPR, shown in more detail in Fig. 10.5.1. The transitman first turns off the 90° angle as best he can with the instrument, and sets the mark R temporarily until he can verify his angle. Verifying the angle is done by meas-

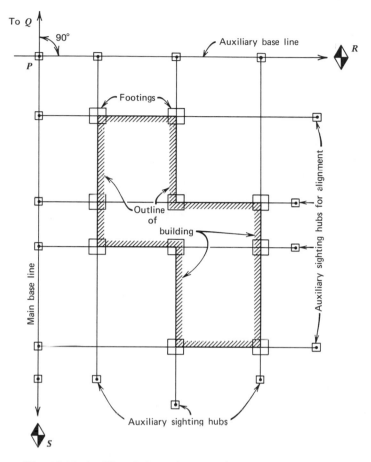

Fig. 10.4.2 Auxiliary hubs and targets for sighting from base lines.

uring it, using the method of repetition, as described in some detail in Section 6.8.

Table 10.5.1 shows a set of notes for six repetitions (three direct, three reverse) for angle *QPR*, revealing that it was set very nearly correct. The angle is seen to be 90°00′08.3″ when the more precise (repetition) measurement is made, and there must be a slight adjustment of the mark at peg *R*. Setting *R* north (counterclockwise) from its initial location will render the angle between the base lines truly a correct right angle. In this instance the shift would seem to be slight, but the small error, if left uncorrected, could magnify to greater proportions and cause closure troubles as the job progressed.

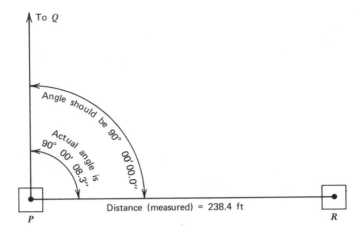

Fig. 10.5.1 Setting the right angle between two base lines.

Table 10.5.1   *Transit Notes for Angle by Repetition*

Hub	Rep.	Tel.	Reading	Angle
Q	0	D	0°00′20″	0°00′00″
R	1	D	(30°00′10″)	(89°00′50″)
R	3	D	270°00′30″	270°00′10″
R	6	R	180°01′10″	180°00′50″

$$(6R) \div 6 = 30°00'08.3''$$
$$\text{Add } (360 \div 6) = 60°$$
$$\text{True} = 90°00'08.3''$$

10.6   Typical Survey Controls for Housing Construction

Figure 10.6.1 shows the minimal controls required to be set by the engineer surveyman in one western state for housing construction. Practices vary from state to state and even within the same state.

10.7   Typical Survey Controls for a Multistory Building

Figure 10.7.1 shows typical minimum control points to be set by the engineering surveyman for construction of a multistory building in the same western state. Again, requirements may vary from place to place.

TYPICAL ENGINEERING SURVEY WORK ON
HOUSING CONSTRUCTION

GRADE AND LOCATION STAKES FOR UTILITIES, SEWERS,
CATCH BASINS, WATER MAINS, ETC. AS MAY BE REQUIRED

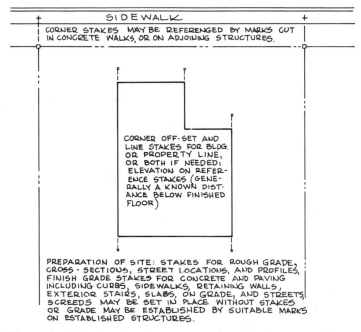

SIDEWALK

CORNER STAKES MAY BE REFERENCED BY MARKS CUT
IN CONCRETE WALKS, OR ON ADJOINING STRUCTURES.

CORNER OFF-SET AND
LINE STAKES FOR BLDG.
OR PROPERTY LINE,
OR BOTH IF NEEDED;
ELEVATION ON REFER-
ENCE STAKES (GENE-
RALLY A KNOWN DIST-
ANCE BELOW FINISHED
FLOOR)

PREPARATION OF SITE: STAKES FOR ROUGH GRADE,
CROSS-SECTIONS, STREET LOCATIONS, AND PROFILES,
FINISH GRADE STAKES FOR CONCRETE AND PAVING
INCLUDING CURBS, SIDEWALKS, RETAINING WALLS,
EXTERIOR STAIRS, SLABS, ON GRADE, AND STREETS;
SCREEDS MAY BE SET IN PLACE WITHOUT STAKES
OR GRADE MAY BE ESTABLISHED BY SUITABLE MARKS
ON ESTABLISHED STRUCTURES.

Fig. 10.6.1 Typical engineering survey work on housing construction.

10.8 Setting Out a Steel-Frame Building

The steel-frame building is set out by aligning the columns along their correct axes. The general outline of the building is initially delineated, in the usual manner, from base lines. Footings or piers for the columns are similarly located from the base lines, as shown in Section 10.4. Batter boards or batter stakes can sometimes be conveniently placed for string alignment of the column footings (as in Fig. 10.3.1) to aid the workmen digging for and setting footing or pier forms.

Vertical-control information is at this juncture made available also so that concrete piers can be poured nearly up to grade to receive the steel base plates. For convenience, column bases are always poured low by $\frac{1}{2}$ to 1 in. or more, and steel shims are placed to bring the column base plates level and up to proper grade. At each column base the surveyman

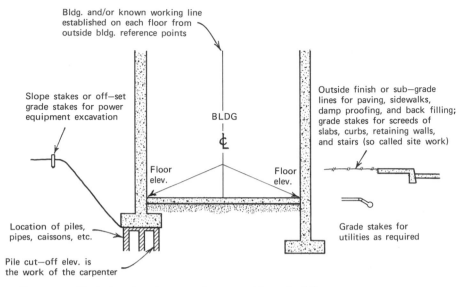

Bldg. and/or known working line established on each floor from outside bldg. reference points

Slope stakes or off—set grade stakes for power equipment excavation

BLDG

Floor elev.

Floor elev.

Outside finish or sub—grade lines for paving, sidewalks, damp proofing, and back filling; grade stakes for screeds of slabs, curbs, retaining walls, and stairs (so called site work)

Location of piles, pipes, caissons, etc.

Grade stakes for utilities as required

Pile cut—off elev. is the work of the carpenter

Fig. 10.7.1 Typical engineering controls on a multistory building.

supplies an elevation mark on the poured footing pier to assure the shimming of the column base plate to its correct vertical height.

Because bolts, poured integrally into the pier, fix the column location horizontally, either a string alignment from batter stakes or a transit sighting is needed to set the (plywood) template that holds the bolts in position. This template is set atop the pier forms and its final alignment in both horizontal directions must be thus assured within the job tolerances from the control base lines. (Column base-plate holes are overlarge to allow some leeway, and bolts are usually amenable to some adjustment laterally, within reasonable tolerances.) Once the base plates for the columns are set, a transit (or string and plumb bob) alignment is best made as a final check before the shims are grouted in place and the columns set.

As the columns are set in place and the framing begins, the structure's alignment and vertically must be verified. Taped measurements between columns of every bay are made, minor corrections being usually possible through hammering or jacking. Plumb wires along columns and/or transit alignment will need to be made to assure verticality, which is maintained by diagonal bracing cables (with turnbuckles for pulling up) until final bolt-tightening or riveting fixes the structural alignment and verticality in final form. For high structures, elaborate extensions of the same procedure are required, as dictated by the nature of the project. Experimental work using vertical lasers has been done for very tall structures, and this is becoming a feasible alignment procedure.

10.9   Setting Out a Reinforced Concrete Building

Placement of the foundation and footings for reinforced concrete structures is done by the same alignment procedures as for steel structures, except that measurements are made to place concrete forms instead of steel base plates. String alignment and transit alignment are both possible and both used, with offset measurements made to the column forms as they are placed and fixed into position. Column reinforcement steel poured into the footings and extending into the columns is aligned in virtually the same manner as are footing bolts for steel columns.

As the building rises, vertical control of building floors is maintained by tape measurements from below. Spirit leveling (a necessity for large-area buildings) is used to place reference marks on poured concrete columns, unstripped forms, or poured concrete floors to carry the levels upward in the structure. Much of this may have to be done up along the outside of the building by taping to avoid interior obstructions. Shaftways frequently afford opportunities to drop tapes for vertical control.

Vertical alignment of columns performed normally by use of the carpenter's level against the formwork may sometimes prove too crude to maintain verticality. Outside sighting with a transit along the face of the building must be maintained to avoid serious misalignment and to introduce corrections when needed. Special brackets for alignment by plumb lines on two axes can be maintained along the face of the building, especially needed for high structures. This type of job is essentially one of successive corrections.

10.10   "As-Built" Measurements

When a building is completed, a certification of its location and the placement of columns, underground utilities, and important interior facilities must be made. "As built" seldom is equivalent to "as planned," so care must be taken to make all needed measurements during construction, altering or annotating the proper drawings for a record. Usually such measurements are made from column axes in the horizontal plane, and from finish-floor levels in the vertical to establish "as built" locations.

The complete set of "as-built" plans is usually required or should be furnished for the proper fulfillment of the job. Proper cooperation between contractor(s) and resident engineer is essential at every stage to ensure this.

11

Trenching and Pipelaying

11.1   Control Marks for Trenching

When digging trenches to install drainage lines or sewer pipes, care to maintain the correct depth so that the water will flow by gravity is essential. Vertical control is of primary importance, more important even than horizontal control for such pipes.

For trench excavations, the center line is marked by stakes driven to the correct horizontal alignment every 100 or 50 ft, although sometimes these marks are placed at a preselected distance offset to the side of the trench opposite where the excavated material is to be cast. Markings on these offset stakes will give the station (e.g., 7 + 50), the offset distance (e.g., 5.0 or 6.0 ft), and the depth of cut from the stake to the bottom of the trench or to the invert of the pipe to be laid in the trench. Horizontal alignment of these stakes is done with the transit; the elevation of the top of the offset hub is measured with the engineer's level since vertical control is so important.

As seen in Fig. 11.1.1, the offset hub can be driven to a correct elevation (if the terrain is favorable), so the top of the hub is an exact number of feet above the invert of pipe. (The invert of the pipe is the inside

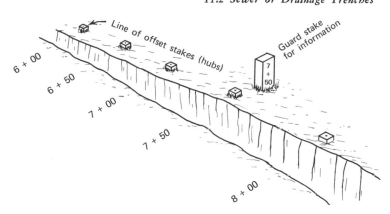

Fig. 11.1.1 Offset subs for trenching, with sample of guard stake.

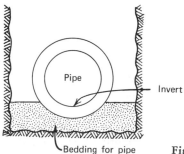

Bedding for pipe   Fig. 11.1.2 Invert of pipe shown.

bottom, where the water flows, as in Fig. 11.1.2.) Or, if this is not done, a mark can later be scribed on the side of the hub at an exact height above the invert. In lieu of either of these, the depth to invert can be written on the guard stake for the top of each hub. The hubs take on the character of "blue tops" then, hubs that carry elevation information and that should not therefore be driven further or disturbed in elevation. "Blue top" is a term stemming from the custom of coloring the top of the hub with lumber crayon. The offset stakes remain to give alignment and grade for the trench and pipe; any original center-line stakes are obviously disturbed and lost when the trenching machine passes through.

11.2 Sewer or Drainage Trenches

In order that drainage or sewer systems operate as designed, all trenches must be excavated true to line and grade. Horizontal alignment is not

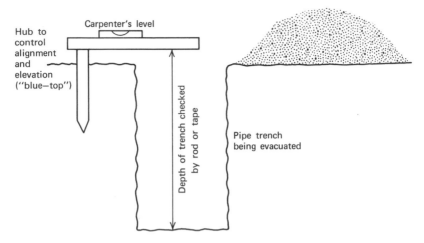

Fig. 11.2.1 Checking depth of trench cut.

very difficult, for a line can be chalked, painted, scratched, or otherwise marked on the surface for the machine operator. More than occasional checking is required to establish that the bottom of the trench is being cut to the right grade. Checking with a board held level from one of the offset grade stakes against a level rod standing in the trench or a suspended tape (Fig. 11.2.1) will quickly check depth. It is usual that the excavation goes a few inches below the bottom of the pipe to allow for a bedding material such as gravel, sand, or crushed rock—so this proper allowance must be made. See Fig. 11.2.2 for typical usage.

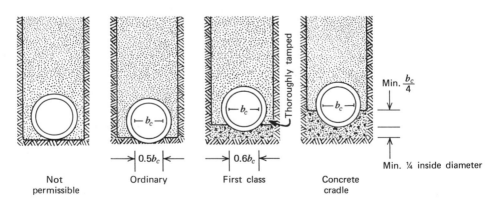

Fig. 11.2.2 Typical requirements for bedding the pipe.

11.3 Batter Boards Over a Trench for Pipelaying

Trenching for laying sewer pipe is frequently a critical task because of relatively low gradients of sewer lines in flat areas and the consequent large pipe sizes needed for slow flow. Preventing too low or too high placement of sewer lines and their associated manholes calls for careful effort and almost invariably for good batter boards. For pipelaying, batter boards are placed to straddle the trench. They are erected quickly after the passage of the trenching machine, usually from the preset grade stakes offset from the trench, as in Fig. 11.3.1. By a string level or a carpenter's level atop a board the grade can be transferred to the batter boards.

The cross board is affixed so as to be level and at a predetermined convenient height above the pipe invert (lowest inside surface, where water flows). Strings are stretched taut between successive batter boards to guide horizontal alignment of pipe, but especially to control vertical placement. (See Fig. 11.3.2.)

By using a templet called a "story pole" or L-shaped rod, workmen can find the invert elevation by the known convenient distance on the story pole from string to invert, and the pipe is placed to exact elevation and bedded in place. (See Fig. 11.3.3.) Occasional verification of any batter-board settlement or disturbance is advisable during the operation, before extensive backfilling, to prevent costly blunders.

It sometimes happens that the sides of the trench are not firm, and batter boards cannot be employed. Even the control hubs may be lost or may have to be set back considerably more than 5 or 6 ft. In such cases, different and more difficult techniques may have to be employed, such as

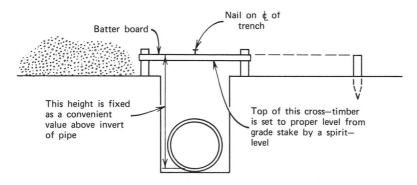

Fig. 11.3.1 Setting a batter board from the grade stake.

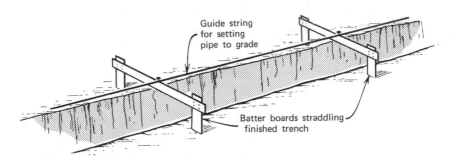

Fig. 11.3.2 Guide strings joining batter boards for pipe alignment.

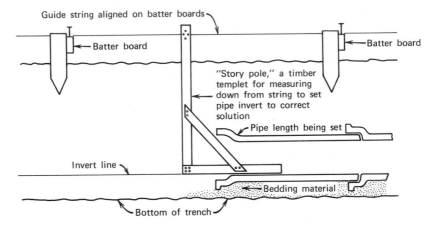

Fig. 11.3.3 Use of a "story pole" or templet for setting pipe to correct depth below batter boards.

using a levelman to stand by and read a rod held on the top or on the invert of each length of pipe as it is being placed. Other methods suggest themselves, among them some that are discussed hereafter.

11.4 Transit to Replace Batter Boards

In lieu of batter boards and string, a sight line may sometimes be set up with an inclined telescope to accomplish the same purpose. In principle, it works the same way, as seen in Fig. 11.4.1.

The transit (or level) is inclined and retained in that position during

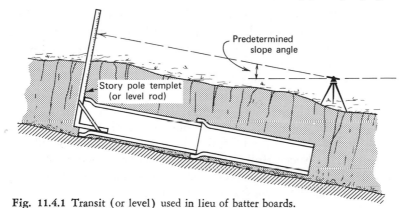

Fig. 11.4.1 Transit (or level) used in lieu of batter boards.

the trenching and pipelaying to avoid the need for batter boards, but it does require an attendant to make frequent sightings. The instrument should be checked frequently to see that it has not settled or been disturbed.

11.5 Laser Beam to Guide Trenching for Pipelaying

In the event that a transit or level telescope is aligned with the trench and inclined upward or downward for pipelaying, as described in Section 11.4, it can also be used to observe a mark placed on the digging machine to control the depth of the trench. This procedure suggests a newer technique, the use of a laser beam. The laser is aligned with the trench and inclined properly a given distance above the pipe invert. Requiring no attendant operator to stand by, the laser is directed along the

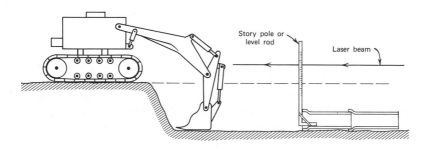

Fig. 11.5.1 Control measurements for sewer-line trenching and pipelaying by laser.

trench to give both alignment and grade for the pipelayers. In addition, the trenching machine uses the beam as a guide for cutting to the correct depth and on the correct alignment, since the beam can be seen where it strikes the machine boom.

The intense red pencil-thick beam shining on the trenching machine's digging arm allows continual, simple, and accurate checking of depth and horizontal alignment without great manpower demand. Figure 11.5.1 shows a backhoe in cutting position, at which instant an observer can easily judge the required depth and signal the operator. The laser beam also serves successfully to set pipe, without batter boards and strings, since an *L*-shaped templet or story pole can be held on the pipe invert to intercept the beam for quick verification.

Lasers guiding bulldozers or trenching machines can be simply a visual guide to the operator, but some machines have even been equipped with light-sensing devices to control the machine itself. Light-sensitive cells pick up the laser beam any time the machine begins to get off grade and operate relays to actuate controls that bring the machine back to grade automatically. The dozer blade or trenching cutter can be maintained easily within a few hundredths of a foot. Such accuracy is not always needed, but a system of this sort can cut down on delays for checking grade and tend to eliminate human error.

11.6   Laser Beams for Pipelaying

As indicated, the laser can serve as a means to set sewer pipe, which is sometimes rather critical because of minimal gradients permissible in flat areas. When the laser can be arranged in a manhole to direct its beam through the pipe being laid in the trench, there will be no interference from the backfilling operation. Such arrangement is possible, in fact necessary, if the pipe is too small for access.

Figure 11.6.1 shows such an arrangement, with the laser braced against the sides and bottom of the manhole. The pipe gradient can be set by means of a "percent grade" dial on the laser support, and can be checked rather easily by ordinary differential leveling after a bit of the trench is dug. In fact, the laser beam will serve well enough even in this situation to guide the trenching machine for line and grade, since it will be interrupted only intermittently as the pipe laying team places its target in the end of each new length being set. A typical target is shown in Fig. 11.6.2, a translucent plastic bulls-eye type set at the correct height above the invert.

In a larger pipe (e.g., 60 or 72 in. diameter of Fig. 11.6.3) one can

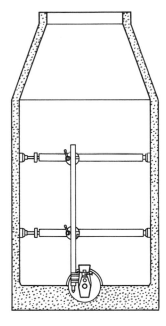

Fig. 11.6.1 Laser in position in a manhole.

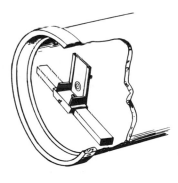

Fig. 11.6.2 Laser target in pipe being set.

place the laser beam near the top of the pipe so that workmen passing through the pipe will interfere less. It is also possible in very large pipes to advance the laser and set it up within the pipe, wedged into position by braces or on a tripod (Fig. 11.6.4) to keep the beam projection shorter.

To assure accurate alignment in either case, a transit is set up on line at the manhole and pointed in the correct forward direction. By drop-

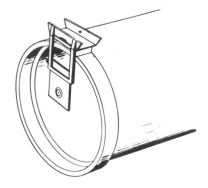

Fig. 11.6.3 Laser target at top of pipe.

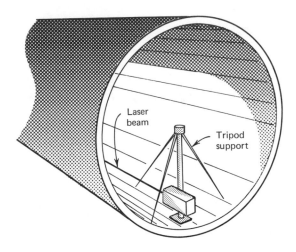

Fig. 11.6.4 Laser supported by tripod in large pipe.

ping a plumb line at any point where the trench is open, the laser beam can be swung to intercept it and achieve proper alignment for the beam. By dropping a tape or a rod down the manhole to the laser beam from the transit's line of sight, and doing the same at the open trench forward, the vertical alignment of the laser beam can thus be verified. It is, of course, simple enough to incline the transit upward to the correct pipe slope as shown on the plan for the sewer line, in which case the laser beam should be inclined the same amount below ground.

Figures 11.6.5 and 11.6.6 are variations that can be arranged to give line and grade by laser for sewers and excavations. In some cases, for

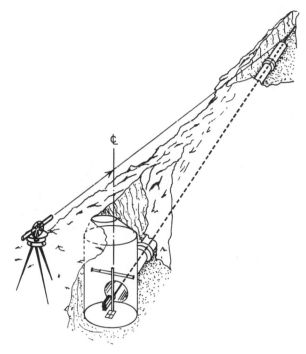

Fig. 11.6.5 Transit to align laser.

example, if the manhole is inaccessible or if the pipe is curved, a laser set above ground may be the answer. It merely means that virtually the same techniques are used but a story pole or rod must be held on the invert or on top of each new length to assure line and grade.

There are several benefits of lasers: no batter boards to set, immediate backfilling possible as the pipe is set, quicker check of line and grade, straighter alignment of pipe, and probably less manpower requirement. Figs. 11.6.7 and 11.6.8 show two different variations of laser instruments used in such alignment work.

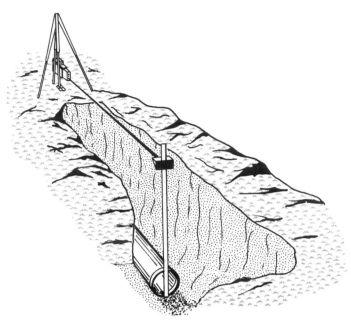

Fig. 11.6.6 Laser above the pipe.

Fig. 11.6.7 Laser mounted on theodolite, with swivel prism to project laser beam through theodolite telescope for automatic alignment. *Courtesy Spectra-Physics.*

Fig. 11.6.8 Alignment laser with "percent grade" dial for beaming other than level in a sewer line.   *Courtesy Spectra-Physics.*

12

Highway Planning and Initial Layout

12.1 Preliminary Highway Map
12.2 Highway Design Map
12.3 Highway Design Sheet Information
12.4 Right-of-Way Boundaries

12.1 Preliminary Highway Map

For planning a highway route, all available maps of whatever scale are useful, especially the topographic maps of the U.S. Geological Survey. These are available at virtually a nominal price through distribution centers or the Map Information Office of the U.S.G.S., Washington, D.C., 20212. The entire United States is mapped at 1:62,500 scale (1 in. = 1 mi, approximately), much is mapped at 1:31,580 (1 in. = 1/2 mi), and a good portion of the nation's urban area is mapped at 1:24,000 (1 in. = 2000 ft).

Working with such maps or others specially prepared, the highway project is narrowed down to a specific routing at an early stage. For accurate "preliminary" planning, a topographic map is then required at a scale of 1 in. = 200 ft (1:2400). Field work involved with producing such a map includes a ground-staked control traverse and a level run for setting out bench marks along the route.

Preliminary planning of highways requires obtaining some ground control, horizontal and vertical, but present-day procedures utilizing topographic mapping by aerial surveying methods make for very little ground work. Plans, profiles, cross sections, and virtually all the design is accomplished in the office with only incidental field measurement or verification.

12.2 Highway Design Map

For design purposes, a map at a scale of 1 in. = 40 ft must be prepared, on which is drawn the details of layout: all tangent and curve information, interchanges and intersections, right-of-way details and fence lines, culverts, bridges, overcrossings and undercrossings, and—most important —all contours and elevation information. When this route alignment plan is drawn, giving the information needed to locate and lay out the route of the highway, the alignment consists of circular curves and tangents (straight portions), as best fit the conditions of the terrain, aesthetics, and economics. Culverts, structures, road intersections and interchanges, and all pertinent construction data are given on the plans by their grid coordinates for every control point, and their route stationing along the center line. From the route alignment plan sheets, the construction is laid out and carried through.

When the contract is awarded for construction, field measurements begin, although it sometimes happens that a preliminary line (P-line) has been staked out under some types of design. Some sort of ground surveying has definitely been done, and some ground control exists. It is evident then that the layout of the construction will be accomplished from the existing control points that will be furnished.

The engineer of location and design makes extensive use of maps worked up from aerial surveys. Lines to be run in the field are laid out first on the maps. When plotted, the position of each ground control point is established and computed, and its location is identified on the ground for the guidance of those entrusted with making the preliminary surveys. Physical characteristics that may influence the establishment of a line or zone of location are carefully evaluated and appraised prior to commencing the detailed ground surveys and layout.

12.3 Highway Design Sheet Information

Specifically, the highway design sheet shows information of immediate use to the surveyman for layout on the ground. Control points of the traverse will be described and listed with N and E coordinates; control bench marks will be described and listed with elevations. Also on the drawings the following highway data will be furnished:

1. Stationing on center line, including beginning and end of project.
2. Coordinates of beginning point, of *P.C.*, *P.I.*, and *P.T.* for each horizontal curve, and also the stationing for each of these. (See Section 13.3).

3. Arc length, angle of intersection (Δ), or radius, and coordinates of center for each horizontal curve.

4. Stationing, coordinates, and elevation(s) for culvert and stream crossings, bridges, underpasses or overpasses, road intersections, and other construction items.

5. Elevations of beginning and ending points, of *PVC*, *PVI*, and *PVT* for the vertical curves, and percent gradient for each tangent. (See Section 14.3).

From such information the work can be laid out and carried through to completion.

12.4   Right-of-Way Boundaries

Right-of-way taking for a new highway involves measurements also, which rightly are construction measurements, but a special word is appropriate here. When property lines are reconstituted or set to form the boundary of a highway right-of-way, the abutting property owners' existing surveys should be properly adjusted to fit the land remaining to them. This is never an easy task, never one that can be lightly undertaken by those inexperienced in the vagaries of property surveys.

To avoid inconsistencies between highway right-of-way descriptions and those of the adjoining properties requires the expertise of the land surveyors locally involved, who are presumably most knowledgeable. Freqently old, inadequate, and even erroneous descriptions of adjacent properties make it necessary to work out new and correct descriptions that will be consistent. The problem exists; it must be faced; it cannot be lightly swept under the rug. It is frequently a problem of working from two different controls, one perhaps better because it is newer and more accurate, but a reconciliation of records must be effected at the time of acquisition of the highway right-of-way. And marking of the boundaries involved should be undertaken at the time of highway construction by use of permanent monuments for the future good of all concerned.

13

Horizontal Circular Curves

13.1 The Circular Curve

Although the bulk of a highway may consist of tangents (straightaways), these must be connected by curved portions of road. For high-speed highways, the curves used are circular curves. While some roads may be designed with spiral curves as the transition to or from the circular curves, these will not be discussed herein. Treatises on horizontal curves exist; here the basics of the circular curve will be simply stated, with a simple treatment of applications consonant with the present work. It should serve as an acquaintance to the terminology and practice.

13.2 Degree of Curve versus Radius

"Degree of curve" is the central angle of the circle that subtends an arc of 100 ft, as Fig. 13.2.1 shows. It is obvious that a sharper curve has a

175

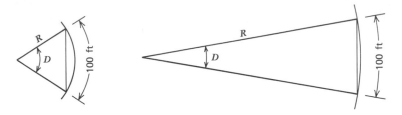

Fig. 13.2.1 Definition of degree of curve: angle subtending 100 ft arc (note that the chord will be less than 100 ft).

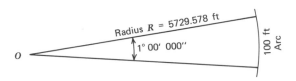

Fig. 13.2.2 Relationships in a 1° curve.

shorter radius and a higher degree of curvature, and a flatter curve has a longer radius and a lower degree of curvature. Either "degree of curvature" or "radius of curve" will describe the particular curve, and both are in common use. The use of "degree of curvature" is currently becoming less popular in highway work in the United States, and the "radius of curvature" concept is more in vogue. In calculations, especially with computing machines, the radius method is easier; in ground layout it is easier and more traditional to set curves by the degree-of-curve concept. For a 1°-curve, as in Fig. 13.2.2, the radius is calculated thus:

$$\frac{100}{2\pi R_1} = \frac{1°}{360°}; \quad R_1 = \frac{18000}{\pi} = 5729.578 \text{ ft}$$

The radius R of any curve is inversely proportional to the degree of curve D; thus,

$$R_2 = \frac{5729.578}{2} = 2864.789; \quad R_3 = \frac{5729.578}{3} = 1909.859, \text{ etc.}$$

Conversely, the degree of a curve can be found from its radius as follows:

$$D_{100} = \frac{5729.578}{100} = 57.2958° \text{ for } R = 100 \text{ ft}$$

$$D_{1000} = \frac{5729.578}{1000} = 5.72958° \text{ for } R = 1000 \text{ ft}$$

$$D_{2000} = \frac{5729.578}{2000} = 2.86479° \text{ for } R = 2000 \text{ ft}$$

$$D_{3000} = \frac{5729.578}{3000} = 1.90986° \text{ for } R = 3000 \text{ ft},$$

and so on.

It is anticipated that the degree of curve may be redefined when the United States moves into the metric system, but a metric degree of curvature would have limited application. European railroad and highway practice has always been based on radius (not degree) of curve; any American railroads would either continue their degree of curve based on feet or convert to radius of curve and use meters, and American highways will probably go to metric radius and abandon degree of curvature. If "degree of curve" were to be redefined when the United States adopts the metric system, probably the "station" would be the 10-m arc-length, and the D_m (metric degree of curve) would be defined as the angle subtending a 10-m arc. This will not be a problem for United States construction for at least a few years, so no effort will be made herein to accommodate to it at this time.

13.3   Circular Curve Theory

The basic circular curve is shown in plan in Fig. 13.3.1, with standard nomenclature as indicated. The word "tangent" in general is used in this context to signify the straight (not curved) portion of a road or route. From Fig. 13.3.1, certain other sometimes useful relations are also seen:

The "long chord" is c:

$$c = AB = AG + GB$$
$$= 2 \sin \tfrac{1}{2}\Delta$$

The "mid-ordinate" is M:

$$M = HG = OH - OG$$
$$= R - R \cos \tfrac{1}{2}\Delta$$

The "external" is E:

$$E = HV = OV - OH$$
$$= R/\cos \tfrac{1}{2}\Delta - R$$

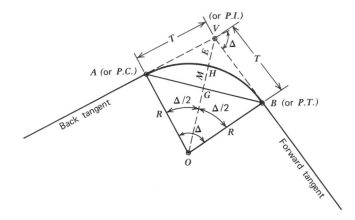

Fig. 13.3.1 Elements and nomenclature of the horizontal circular curve.

Δ = angle of intersection
R = radius of curve
P.I. = point of intersection (V)

P.C. = point of curvature (A)
P.T. = point of tangency (B)

A most used relationship, however, is the "tangent" (or "semitangent") of the curve, which is T:

$$T = AV \quad \text{and also} \quad T = BV; \; T = R \tan \tfrac{1}{2}\Delta$$

13.4   Calculation of Circular Curve Stationing

Stationing of the route is carried around the curve, not through the tangents to the vertex (*P.I.*), and consequently the value of T becomes important. The "length of curve" L from *P.C.* to *P.T.* is a function of the intersection angle (Δ) and of the radius (or degree) of the curve. It is calculable simply as the number of times that a 100-ft arc will fit in between the *P.C.* and the *P.T.*—the same as the number of times that the degree-of-curve D will fit into Δ. Thus, from Fig. 13.4.1,

$$L \text{ (in Sta.)} = \frac{\Delta}{D}$$

If $\Delta = 21°15'30''$
and $R = 1500$ ft

$$D = \frac{5729.578}{1500} = 3.81972°$$

$$L \text{ (Sta.)} = \frac{21.25833°}{3.81972°} = 5.5654$$

or $5 + 56.54$

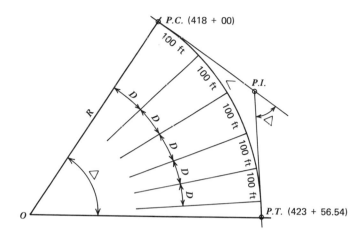

Fig. 13.4.1 Evaluation of length of curve.

If, in this instance, the *P.C.* were at Sta. $418 + 00.00$, then the station of the *P.T.* would be $423 + 56.54$; but it never quite happens that the *P.C.* will fall exactly at a full station. But neither is this a disadvantage, as will be seen. Fig. 13.4.2 demonstrates the calculations involving stationing:

$T = R \tan \frac{1}{2}\Delta$

$\quad = 2000 \tan 14°20'15''$

$\quad = 2000 \ (0.2555940)$

$\quad = 511.188 \text{ ft}$

N (*P.I.*): $914 + 48.700$

$-T = \qquad 5 + 11.188$

A (*P.C.*): $909 + 37.512$

$+L = \qquad 10 + 00.945$

B (*P.T.*): $919 + 38.457$

$\Delta = 28°40'30'' = 28°40.5' = 28.6750°$

$$D = \frac{5729.578}{2000} = 2.86479°$$

$$L = \frac{\Delta}{D} = \frac{28.6750}{2.86479} = 1000.945$$

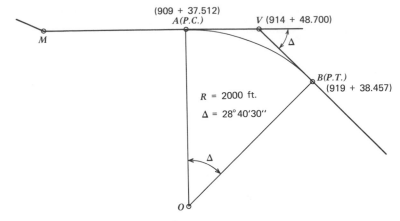

Fig. 13.4.2 Stationing the curve points.

Note that although the *P.I.* would seem to have a station number (914 + 48.700), stationing of the route proceeds through the curves and not through the vertex; the *P.I.* stationing therefore is virtually useless.

The simple method of laying out a short-radius curve is to tape from the center, whenever the center can be located and when taping can be done without encountering obstructions. Usually, however, when the curve radius exceeds about 100 or 150 ft, there are difficulties with this type of endeavor; trees, buildings, earthbanks, and so on, get in the way. Layout of large circular curves is regularly done rather with transit and tape from the *P.C.* or the *P.T.*, or even from any point on the curve, once some simple principles are known. (Complex problems of curve design and layout are treated more fully in special treatises on curves.)

13.5 Calculating the Deflection Angles for Setting the Curve

Here is an illustrative example, assuming a 5° curve with the *P.C.* at Sta. 37 + 20.16 and an intersection angle (Δ) of 22°00′ exactly. The necessary calculations for obtaining the deflection angles are shown; it is these deflections from the back tangent that are the secret to laying out this curve on the ground. Figure 13.5.1 illustrates this computation.

$$P.C.\ (A)\ \text{at Sta. } 37 + 20.16$$

$$D = 5°00'$$

$$L = \frac{22.0000}{5.0000} = 4 + 40.00$$

$$(440.00\ \text{ft})$$

$$P.T. \ (B) \text{ at Sta. } 41 + 60.16$$

The value of d_1 must be first calculated:

$$\frac{d_1}{D} = \frac{79.84}{100.00}$$

(since 79.84 is arc distance from *P.C.* to Sta. 38 + 00)

$$\tfrac{1}{2} \, d_1 \text{ (in min. of arc)} = \frac{\tfrac{1}{2}\,D}{60} \frac{79.84}{100.00} = 0.3\,D\,(79.84)$$

$$= (0.3)(79.84)(5.00) = 119.76' = 1°59.76'$$

Thus, $\tfrac{1}{2} \, d_1$ is seen to be the initial deflection from the back tangent that will be used to set out the hub for Sta. 38 + 00. This deflection angle, formed by a tangent and a chord, is measured by half the central angle. Note that d_1 would be 3°59.52′, and so $\tfrac{1}{2} \, d_1$ is 1°59.76′, as shown.

The other deflection angles (all from tangent *AV*) are calculated for each full station on the curve as in Table 13.5.1, and field notes are worked up for convenient layout. In the calculation, note that the total deflection angle for the *P.T.* should work out to be half of the intersection angle, which is also half of the central angle for the entire curve. The principle is one of geometry; angle formed by two chords, or by a tangent and a chord, is equal to one-half of the intercepted arc. Since the full arc is 22°00′, the final total deflection angle is 11°00′ in this case. Note that the total deflection to Sta. 39 + 00 is one-half $(d_1 + D)$, being an angle formed by a tangent and a chord, or 1°59.76′ + 2°30.00′ = 4°29.76′ in this case. Similarly each total deflection angle is related to the arc it intercepts, being half of the arc it intercepts or half of the central angle subtended by that arc.

Table 13.5.1   Field Notes for Circular Curve Layout

Sta.	Point	Total Deflection	Calculation
37 + 20.16	*P.C.*	0°00.00′	$\tfrac{1}{2}d_1 = 0.3\,(79.84)\,(5.00) = 1°59.76'$
38 + 00		1°59.76′	Each increment to the initial deflection angle
39 + 00		4°29.76′	can be seen to be simply $D/2$ (=2°30′) for
40 + 00		6°59.76′	each full 100-ft station.
41 + 00		9°29.76′	
41 + 60.16	*P.T.*	11°00.00′	$\tfrac{1}{2}d_2 = 0.3\,(60.16)\,(5.00) = 90.24 = 1°30.24,$

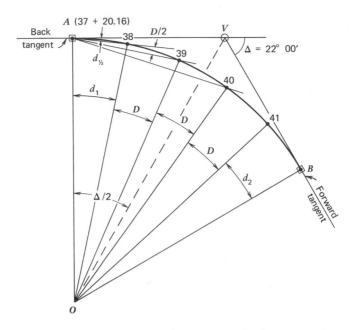

Fig. 13.5.1 Deflection angles for horizontal circular curve.

13.6 Setting Out the Curve from the P.C.

To set out the full stations on the curve from the notes of Table 13.5.1, the *P.C.* is occupied by the transit, and the *P.I.* sighted with 0°00.00′ as the backsight reading. Then the first total deflection angle (1°59.76′) is turned and the chord distance to the first full station (Sta. 38 + 00) is laid off from the *P.C.*, and a hub set. Figure 13.6.1 shows this procedure. To set out the next full station (Sta. 39 + 00.00), the total deflection at *P.C.* to that station (4°29.76′) is turned from *P.I.*, but the distance measurement is made by two tapemen along the chord from the previous full station (Sta. 38 + 00) to the one being set. This procedure involves an unusual and relatively weak intersection, calling for swinging the tape end until it crosses the line of sight and there setting a point. The next full station (Sta. 40 + 00) is set by turning at *P.C.* the proper total deflection from *P.I.* (6°59.76′) and measuring the proper distance from its preceding station (Sta. 39 + 00).

This process continues until the *P.T.* is set. Normally, errors accumulate in this rather weak method and cause the *P.T.* thus set to fall at other than its correct location. So, the *P.T.* is always best set beforehand

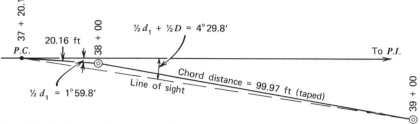

Fig. 13.6.1 Setting out the curve stations.

by sighting along from the *P.I.* of the curve towards the next *P.I.*, that is, along the forward tangent, and by taping along from the *P.I.* the proper tangent distance *T*. It is usual, then, that both the *P.C.* and the *P.T.* are set at the same time, when the *P.I.* is first occupied, by taping the distance *T* from the *P.I.* along each tangent in turn.

13.7 Setting Curve Points from Anywhere on Curve

Sometimes it is impractical or impossible to set the entire curve by turning deflection angles from the *P.C.*, for some of the sights get to be rather long. So, the transit is moved up along to some other point along the curve to continue the work (even moved to the *P.T.*). Because of the geometric principles involved, the same computed total deflection angles can be used, however. Setting up at any point on the curve, one orients the transit by sighting the *P.C.* or the *P.T.* or any other point on the curve by this rule:

> To set a curve point from any point on the curve, backsight to any other curve point with the deflection set on the transit which was used to set the point sighted. Then it is possible to use the calculated total deflection angles to set other curve points, assuming distances are measured properly from adjacent stations.

To set points on the curve, then, move up on the curve to any point and backsight to any other point on the curve, setting the vernier of the instrument to the total deflection angle of the station being sighted. One may thus use the notes already computed for total deflections. For example, if after setting Stas. 38 and 39, a further sighting is blocked by some obstruction, the transit can be moved forward to Sta. 39 and set up on that point. This is shown in Fig. 13.7.1. Then one would backsight to

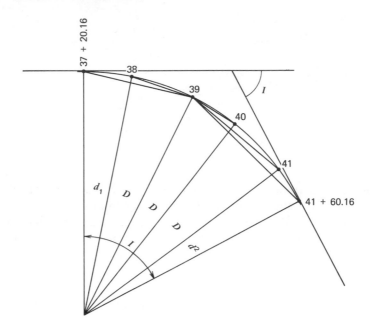

Fig. 13.7.1 Setting curve points from anywhere on the circle.

the *P.C.* with 0°00.0′ set on the instrument (the value of deflection angle that was used to set in the *P.C.*), and be in a position to set in any other point on the curve. One would verify that a reading of 1°59.8′ cuts the hub at Sta. 38 (the value that was used to set that station point), then plunge the transit and sight forward to set in Sta. 40 by using its deflection angle (6°59.8′), taping the proper chord length (99.97 ft) from Sta. 39. One would then set in Sta. 41 by using its deflection angle (9°29.8′), taping the proper chord length (99.97 ft) from Sta. 40. Finally, one would set in the *P.T.*, or check it, by using its deflection angle (11°00′), taping the proper chord length (60.16 ft) from Sta. 41.

In other words, no new deflection angles would need to be tabulated; those already calculated would serve, no matter where on the curve the transit is set up. In fact, the method works if the transit is set up on the *P.T.*, or on a half-station, or on any point that is known to be on the curve (even if its stationing is not known). Several other curve-staking methods exist, though none is better than the deflection-angle method. In some circumstances one may wish to use the chord-offset method, or the tangent-offset method, in which case it is necessary to refer to a more specialized book on the subject.

13.8 Chord versus Arc Distances

Distances used in the preceding layout method must, of course, be chord distances (not arc distances), even though the calculation of the deflection angles is done by arc distances. For the usual situation the arc and the chord are not greatly different, but they cannot be used interchangeably for accurate work, especially for sharp (short-radius) curves.

From the known properties there can be established a relationship between arc length, a and chord length, c:

$$\frac{a}{2\pi R} = \frac{d}{360} \qquad\qquad \frac{c}{2} = R \sin \frac{d}{2}$$

$$R = \frac{360}{2\pi d} \cdot a \qquad\qquad c = 2R \sin \frac{d}{2}$$

Combining,

$$c = \frac{360}{\pi d} \cdot a \sin \frac{d}{2} \text{ (general)}$$

or when $a = 100$ ft and $d = D$ (degree of curve)

$$c_{100} = \frac{36000}{\pi D} \sin \frac{D}{2} \text{ (full station)}$$

In practice, tables exist, or short approximate formulas are used. In Tables 13.8.1 and 13.8.2, some handy values are tabulated, principally to give some notion of the relative differences between chord and arc for some combinations.

Table 13.8.1 *Comparison of Chords and Arcs of Circular Curves for Various Degrees of Curve*

Degree of Curve	Chord for 100-ft Arc	Chord for 50-ft Arc	Chord for 25-ft Arc
1	100.00	50.00	25.00
2	100.00	50.00	25.00
3	99.99	50.00	25.00
4	99.98	50.00	25.00
5	99.97	50.00	25.00
6	99.95	50.00	25.00
7	99.94	50.00	25.00
8	99.92	49.99	25.00
9	99.90	49.99	25.00
10	99.88	49.98	25.00
11	99.85	49.98	25.00
12	99.82	49.98	25.00

Table 13.8.1—Continued

Degree of Curve	Chord for 100-ft Arc	Chord for 50-ft Arc	Chord for 25-ft Arc
13	99.79	49.97	25.00
14	99.75	49.97	25.00
15	99.72	49.96	25.00
20	99.49	49.94	24.99
25	99.21	49.90	24.99
30	98.86	49.86	24.98
35	98.45	49.80	24.97
40	97.98	49.74	24.97
45	97.45	49.67	24.96
50	96.86	49.60	24.95

Table 13.8.2   Comparison of Chords and Arcs for Various Even-Radius Circular Curves

Radius of Curve (ft)	Chord for 100-ft Arc	Chord for 50-ft Arc	Chord for 25-ft Arc	Chord for 10-ft Arc
2000 and over	100.00	50.00	25.00	10.00
1900	100.00	50.00	25.00	10.00
1800	99.99	50.00	25.00	10.00
1700	99.99	50.00	25.00	10.00
1600	99.98	50.00	25.00	10.00
1500	99.98	50.00	25.00	10.00
1400	99.98	50.00	25.00	10.00
1300	99.97	50.00	25.00	10.00
1200	99.97	50.00	25.00	10.00
1100	99.96	50.00	25.00	10.00
1000	99.96	50.00	25.00	10.00
900	99.95	50.00	25.00	10.00
800	99.93	50.00	25.00	10.00
700	99.92	49.99	25.00	10.00
600	99.89	49.99	25.00	10.00
500	99.83	49.98	25.00	10.00
400	99.74	49.97	25.00	10.00
300	99.54	49.94	24.99	10.00
250	99.34	49.92	24.99	10.00
200	98.96	49.87	24.98	10.00
180	98.72	49.84	24.98	10.00
160	98.38	49.80	24.97	10.00
140	97.88	49.74	24.96	10.00
120	97.13	49.64	24.96	10.00
100	95.89	49.48	24.93	10.00
80	93.62	49.19	24.90	10.00
70	91.73	48.94	24.87	9.99
60	88.82	48.57	24.82	9.99
50	84.15	47.94	24.74	9.98

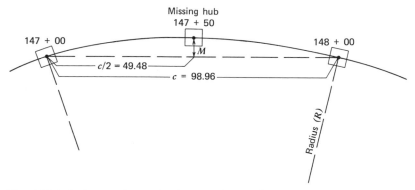

Fig. 13.9.1 Setting a point on circular curve by use of tape only.

Example: If $R = 200$ ft and $C = 98.96$ ft

$$M = \frac{C^2}{8\,R} = \frac{98.96^2}{8\,(200)} = 6.12 \text{ ft}$$

13.9 Setting Curve Points with a Tape

In the event that a missing center-line hub needs to be quickly reset between two adjacent hubs, an approximate method requiring no more than a tape can be used. Figure 13.9.1 shows the half-station being set at Sta. $147 + 50$ on a 200-ft radius curve. The offset M is calculated nearly correctly as $M = c^2/8\,R$, the letter M indicating "midordinate" of the curve in this case. Then a tape (or cord) is stretched between existing hubs and at the halfway mark the offset is measured off 6.12 ft and the missing hub is set. This method will be sufficiently accurate for all practical purposes.

13.10 Setting Curve Points with an Angle Prism

By using a property of the circle, namely that any angle inscribed in a semicircle is a right angle, one could manage also by using only a hand-held right-angle prism to set himself on the circle at any point. It would, of course, require some preparation; the two points he needs to sight, the ends of any diameter, will have to be visible and marked. So, in general, this would apply only to short-radius curves.

To illustrate, assume that a curve of 150-ft radius is to be set at an intersection of two roads, as Fig. 13.10.1 shows. The *P.C.* on Able Road is set by taping back 150 ft from Baker Road, the transit (or even a right-angle prism) can then turn off the 90° desired to set point B, which is $2\,R$ distant (300 ft). From then on, a person carrying a right-angle prism

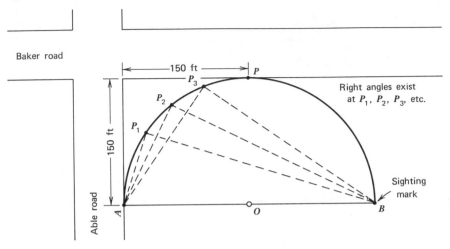

Fig. 13.10.1 Finding points on a circle with a 90° prism.

can set any number of points along the curve between *P.C.* and *P.T.* By merely moving and sighting and guiding himself by observing *A* and *B* simultaneously, he can ascertain when he is on the curve because *A* and *B* will appear superimposed only then. At any point on the curve the right-angle prism sights the ends of the diameter.

14

Vertical Parabolic Curves

14.1   Vertical Alignment and the Parabolic Curve

Vertical alignment of the highway is established by the design engineers in percent gradient (or "grade"), a rise or fall in feet per station. Change of gradient is accomplished by use of parabolic curves, principally because they can be computed conveniently. The results of vertical curve computation appear as elevations.

Theoretically the parabolic curve is ideally suited to horizontal curves on high-speed highways and railroads, although mainly because of tradition, and also because of the difficulty of handling the computations, parabolas have not previously been used for horizontal curves. The theory developed in this chapter is applicable to horizontal as well as to vertical curves, and we may soon find the parabola applied equally to both on high-speed routes.

14.2   Vertical Curve Theory

A simple vertical curve computation can be effected by knowledge of three properties of the parabola. The first is as shown in Fig. 14.2.1:

a. The line joining the midpoint C of a chord AB of a parabola with the intersection D of the tangents at the ends of the chords is bisected by the parabola itself, E. Thus, $DE = EC$.

Figure 14.2.2 suggests the possibility of its use in highway or street work, and the following simple example will show the straightforward computation; the usual nomenclature is used, or

$P.V.C.$ = Point (or beginning) of vertical curve
$P.V.I.$ = Point of vertical intersection
$P.V.T.$ = Point of vertical tangency (end of vertical curve)

In this case, Points A, B, C, D, and E are related to those in the curves of Fig. 14.2.1, so DE will again be equal to EC. By inspection of the curve, it is seen to be 600 ft (or 6 stations) in length, and must connect a $+4.00\%$ grade with a -2.00% grade. As is customary, this is an "equal-station" vertical curve, with 3 station-lengths before and 3 following the $P.V.I.$ The elevation of the $P.V.I.$ is shown here as 100.00 ft to simplify

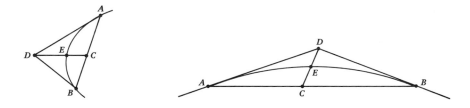

Fig. 14.2.1 First property of the parabola.

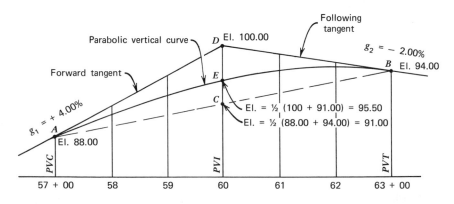

Fig. 14.2.2 The parabolic vertical curve, simple computation.

the illustration; thus, the elevation of each of the tangent points is conveniently found. We need in the present case merely to know the elevations of points A and B, since we want the mean of these two elevations (which will be the elevation of point C). Once we have the elevation of C and of D, we find, from the first property of the parabola, that the parabola lies midway between points C and D. Thus, point E must have an elevation that is the mean of the elevations of C and of D, or 95.50 ft. The computations can be easily followed on Fig. 14.2.2.

14.3   Elevation of Vertical Curve Points

Thus, one elevation, at the midpoint, is known for the curve (95.50), since point E is midway between elevation 100.000 and elevation 91.00 and is on the curve. Another helpful property will next be employed, as shown in Fig. 14.3.1:

b. Offsets from the tangent to the parabola vary as the square of the distance from the point of tangency.

Figure 14.3.2 shows the use of this property in the illustrative example, where one offset is already known (central offset at Sta. 60 + 00, DE = 4.50) and others must be determined (at Stas. 58, 59, 61, and 62). The entire vertical curve shown in Fig. 14.3.3 can now be tabulated in note form, shown in Table 14.3.1.

Table 14.3.1   Notes of the Vertical Parabolic Curve

Station		Tangent Elevation	Offset	Curve Elevation	
55 + 00		80.00			
56 + 00		84.00			
57 + 00	P.V.C.	88.00	0.00	88.00	
58 + 00		92.00	−0.50	91.50	
59 + 00		96.00	−2.00	94.00	
60 + 00	P.V.I.	100.00	−4.50	95.50	Vertical Curve (length = 600 ft)
61 + 00		98.00	−2.00	96.00	
62 + 00		96.00	−0.50	95.50	
63 + 00	P.V.T.	94.00	0.00	94.00	
64 + 00		92.00			
65 + 00		90.00			

Elevations of any other desired points on the vertical curve can be found by using the preceding principles, for example, the half-stations (57 + 50, 58 + 50, etc.), or other intermediate points for which field computation may be needed.

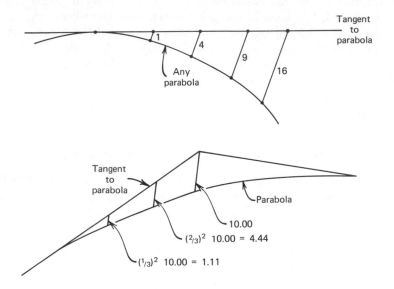

Fig. 14.3.1 Second property of the parabola.

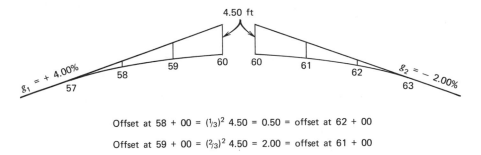

Offset at 58 + 00 = $(1/3)^2$ 4.50 = 0.50 = offset at 62 + 00

Offset at 59 + 00 = $(2/3)^2$ 4.50 = 2.00 = offset at 61 + 00

Fig. 14.3.2 Finding offsets at any point for the vertical parabolic curve.

14.4 High Point or Low Point

Inspection of Fig. 14.3.3 or of Table 14.3.1 shows that the high point (summit) of this curve is at Sta. 61 + 00, by symmetry; it would be the low point (sag) if the geometry of the curve were inverted—with a minus gradient tangent meeting a plus gradient tangent. The high point or the low point is sometimes referred to as the "turning point" of the

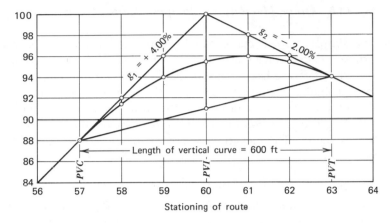

Fig. 14.3.3 The full parabolic vertical curve.

vertical curve. Since it cannot usually be discovered so easily by inspection, the location of the high point or low point on the curve is found as the point of zero gradient ($g = 0\%$) on the curve. Another property of the parabola is helpful here to understand this:

    c. The rate of change of curvature of the parabola varies directly as the distance.

    Thus, the initial gradient changes uniformly with the distance: the rate of change of gradient in percent per station (r) is constant between the initial gradient (g_1) and the final gradient (g_2). In our case,

$$r = \frac{g_2 - g_1}{L \text{ (in Sta.)}} = \frac{2.00\% - (+4.00\%)}{6 \text{ Sta.}} = -1.00\%/\text{Sta.}$$

The initial gradient ($+4.00\%$) will run out to 0% at this constant rate in four stations, giving the high point at Sta. $61 + 00$.

$$g_1 + r(n') = 0\%; \quad n' = \frac{-g_1}{r} = \frac{-4.00\%}{-1.00\%} = 4.00 \text{ Sta. from } P.V.C.$$

where n' is the number of stations needed to run out the initial gradient to zero. Hence we calculate that Sta. $61 + 00$ is the turning point of the sample curve. Its elevation, were it not already known, would be found by the method of Section 14.3.

14.5   Other Methods to Compute Vertical Curves

While any simple vertical curve can be computed from the foregoing, such as those needed for street intersections, highway cross sections, or random needs, there are handbooks available for more unusual problems. One approach that will prove surprisingly simple is the direct use of the quadratic equation for the parabola. In the form given here, it will lend itself to finding the elevation of any desired point on the vertical curve with little difficulty. The elevation of any point along the curve, E_x, is given by:

$$E_x = E_o + g_1 x + \tfrac{1}{2} r x^2$$

where x is the distance in stations from the *P.V.C.*, E_o is the elevation of the curve at the *P.V.C.*, and $r = (g_2 - g_1)/L$, as before. This formula can be solved by hand, by desk calculator, or by electronic computer, and is recommended as a handy means to get random curve elevations as required for staking out. Typically, some points on the foregoing sample curve are here calculated:

Sta.	g_1	x	r	E_o	$g_1 x$	$\tfrac{1}{2} r x^2$	E_x
59 + 00	+4.00	2.00	−1.00	88.00	+8.00	−2.00	94.00
59 + 50	+4.00	2.50	−1.00	88.00	+10.00	−3.12	94.88
60 + 00	+4.00	3.00	−1.00	88.00	+12.00	−4.50	95.50
61 + 70	+4.00	4.70	−1.00	88.00	+18.80	−11.04	95.76

There is also the chord-gradient method, covered in some curve treatises, that will not be described here, since it is not well adapted to field calculation. It is usable, however, for office design of vertical curves. What is contained herein should suffice to work out any field computations that may arise during construction layout.

15

Staking Layout of Highways

15.1   Setting Out Highways

The initial point for a road layout is found from knowing its coordinates from the plan and working to it from the control traverse points that are furnished and can be found nearby. In setting the initial point, the direction of a line will be available, so the roadway direction can be set by properly turning an angle. Constant reference ties to other traverse control points whenever they are available along the route layout is necessary to assure that the roadway be held in proper alignment.

Locations of structures, culverts, and special features may be needed at once, so that work on these can begin as soon as possible. These are set out by their coordinates from the nearest horizontal and vertical control points available from the preliminary work. These procedures are separately described, and the general process of setting a point from known coordinates was covered in Section 8.3.

15.2   Center-Line Hubs

Many center-line hubs will already have been set, in conjunction with work on culverts, bridges, overpasses, and so on, but at this time all the full-station hubs are to be set. It is sometimes better to set in the *P.I.* hubs (see Section 13.3) first, working from different spots along the control traverse or triangulation network. In this way, the curves can be set, and the entire work can be set out more rapidly, with a greater number of field parties. It will also tend to avoid the buildup of errors that could occur along the route of construction if the laying out was a cumulative procedure. But finally, the full-station hubs are put into the ground along the center line.

Sometimes it occurs that a preliminary traverse (*P*-line) has not been set in the field conveniently nearby. Then the procedure is to stake out by traverse (Section 8.2) and by inverse (Section 8.3), the principal alignment points, that is, points of intersection (*P.I.*) and other points on the tangent (straight) sections of the route. From the *P.I.* hub of each curve, the *P.C.* and the *P.T.* hubs are set (and, as noted, these will not ordinarily fall on a full- or half-station peg). Nevertheless, it is from the *P.I.* points and the others set on the ground that each full station is staked out in its proper position.

"Stations" are usually 100 ft apart and United States procedure is to use this 100-ft "station length" as the standard for layout, for earthwork computation, and for all design computation and records. They are seen on plans as Stas. 0+00, 1+00, 2+00, and so on, throughout the project. Discussion is already underway as to what should be the "station length" when the United States goes metric. Some authorities speak of a 20-m length, but this may only lead to confusion; the best method may be to use the 100-m length, but set half-stations and quarter-stations. Time will tell.

By the proper procedure, the center-line hubs are set around each curve, with the stationing progressing through the curves and not through the tangents, as Section 13.4 illustrates. These center-line stakes, on tangents as well as curves, are set to permit the leveling party to obtain ground elevations at the center line of each station, and to obtain cross-section information to right and left. In doing this level work, a record is made of the elevation of the center-line peg at each station—a sort of temporary bench mark from which later to make further elevation determinations for cut and fill. Conveniently, at this time, frequent and fairly permanent bench marks are set outside the construction area as work progresses. The early center-line leveling information may be needed for the roadway design, but normally is only for verification, since modern practice is to have already obtained all this information by photogrammetric (aerial survey) methods by this stage. It can be presumed that both center-line elevations and slope-stake information is already obtainable from the design drawings.

Next there is indicated on the center-line station stakes the depth of cut or fill (in feet and tenths) for the grading operation. Since these center-line stakes will necessarily be removed or covered in the process of construction, it is expected they will have to be reset at least once again at a new elevation, and the process of leveling repeated at least once more.

15.3 Slope Stakes

At this initial pass, side stakes called slope stakes are also set at the edge of construction—where the cut banks and fill slopes will intercept the ground. These are shown in Fig. 15.3.1 for a fill (embankment). These slopes stakes are marked to indicate their distance from the center line and their height above or below the ground at the center-line stakes. Whether for fill or cut, the slope stakes (or edge stakes) must be set by measurement from the center-line hubs. These slope stakes are set about 1 ft beyond their correct location so they will furnish information to the earthwork machine operators without being disturbed. The top of cut-bank or the toe of slope is found, if need be, by computation: $d = \frac{1}{2}$ (subgrade width) + (slope ratio)(difference of elevation between grade at centerline and ground at top or toe).

This is shown by Fig. 15.3.1 as

$$d_L = w/2 + S_L(h_L)$$

$$d_R = w/2 + S_R(h_R)$$

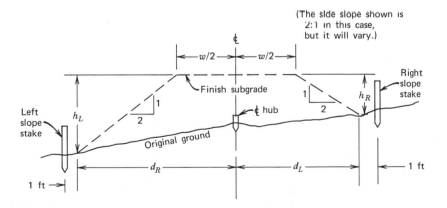

Fig. 15.3.1 Typical roadway cross section (fill or embankment).

Setting a slope stake in the field is a cut-and-try proposition, since the spot must be found where the new slope of the cut or of the fill will intercept the existing ground slope. From the center-line hub a distance is measured out to a tentative slope stake location, and an elevation is taken of the ground at the tentative spot. Then, knowing the values of $w/2$, of S_L (or S_R), and of h_L (or h_R), the theoretical distance d_L (or d_R) is computed and compared with the measured d_L (or d_R). Normally an adjustment will have to be made, a new location tried, and a new spot found. It requires practice and skill, but a spot can be found where the equation is satisfied, the spot where the new slope will meet the existing ground slope.

A record of these slope stakes would be developed by this process, slope-stake notes as shown in Fig. 15.3.2, for example. It is from these

Sta.	L	₵	R	
19 + 00	$\dfrac{-10.7}{23.0}$	$\dfrac{-10.8}{0}$	$\dfrac{-10.1}{22.1}$	Cutbank (+)
				Base 20 ft
				Slope 1:1
18 + 50	$\dfrac{-7.6}{18.4}$	$\dfrac{-7.9}{0}$	$\dfrac{-8.8}{19.4}$	Fill (−)
				Base 14 ft
18 + 00	$\dfrac{-6.4}{16.6}$	$\dfrac{-5.3}{0}$	$\dfrac{-8.6}{12.4}$	Slope 1½:1
17 + 50	$\dfrac{-8.6}{12.4}$	$\dfrac{-7.1}{0}$	$\dfrac{-6.4}{16.6}$	
17 + 00	$\dfrac{-2.6}{10.9}$	$\dfrac{-4.9}{0}$	$\dfrac{-6.0}{16.0}$	
16 + 94.2	$\dfrac{0.0}{7.0}$	$\dfrac{-2.3}{0}$	$\dfrac{-4.2}{13.3}$	
16 + 86.5	$\dfrac{+1.9}{11.9}$	$\dfrac{0.0}{0}$	$\dfrac{-2.2}{10.3}$	
16 + 77.0	$\dfrac{+3.2}{13.2}$	$\dfrac{+1.1}{0}$	$\dfrac{0.0}{10.0}$	
16 + 50	$\dfrac{+6.1}{16.1}$	$\dfrac{+3.9}{0}$	$\dfrac{+4.3}{14.3}$	
16 + 00	$\dfrac{+7.7}{17.7}$	$\dfrac{+5.2}{0}$	$\dfrac{+3.2}{13.2}$	

Fig. 15.3.2 Cross-section notes resulting from the slope-staking procedure (three-level sections).

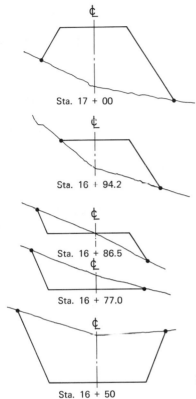

Sta. 17 + 00

Sta. 16 + 94.2

Sta. 16 + 86.5

Sta. 16 + 77.0

Sta. 16 + 50

Fig. 15.3.3 Cross sections drawn for typical cases shown in the slope-staking notes of Fig. 15.3.2.

notes that a set of cross-section drawings such as in Fig. 15.3.3 would be made for volume calculations (as in Chapter 20). Here, however, we wish simply to establish that the slope stakes are a construction guide. Since these slope stakes on each side delineate the limit of cut or fill, then, as stated, they are set 1 ft beyond their proper positions so as not to be wiped out by earth-moving equipment. Needed information written on stakes will help to guide the equipment operators and also to keep handy the correct elevations and alignment of the road during the con-fusion of construction. The slope stakes set at this time must carry the station numbering, the depth of cut or fill to the roadway center line, and the distance to the center line. The constructors will have virtually noth-ing more than these stakes to guide themselves during the entire earth-moving operation.

In a modern procedure where a design drawing is based on a large scale map (1 in. = 40 ft or 1 in. = 50 ft) of the highway, it is virtually

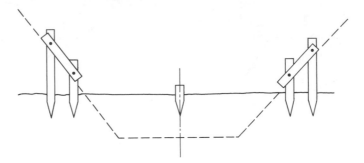

Fig. 15.3.4 Sighting stakes as an alternative to slope stakes.

unnecessary to set out slope stakes by the field method described. The edge of cut or toe of slope is shown on the design plan drawing, and the distance to the slope stake can be scaled accurately enough to set the stake. The ground elevation might be checked when the stake is driven—at least in some random cases. However, the cost of setting slope stakes to pinpoint accuracy could easily be higher than the cost of slight earthwork adjustments by machine during the progress of the work.

It would be possible and perhaps helpful to set cut and fill sighting stakes in place of, or perhaps to supplement, slope stakes in special cases. This may be helpful in digging canals or building dikes (levees), although it is not a usual method. Figure 15.3.4 shows such sighting stakes in place for a cut section of roadway or canal.

15.4 Reference Staking

It is sometimes necessary to set auxiliary stakes for deep cuts and high fills because slope stakes then tend to be too far removed from the working surface for guidance to equipment operators. Frequent realignment by transit and resetting of temporary bench marks is called for in any such difficult area.

A practice of marking the tops of driven pegs along the way with blue lumber crayon ("blue tops") serves to indicate that these carry an elevation and ought not to be driven further or disturbed. These can be set at the edge of construction early in the work—at the slope stakes—and only later, at the center line as finished elevation of the subgrade is attained.

Since the purpose of the center-line stakes will soon have been served and they will be covered or torn up by the first pass of equipment, they are obviously no longer needed. There will be need later on, however, to set them once again in place when the subgrade has been reached and the center line of the roadway must be reestablished for the final shaping of the subgrade and the paving operation. It is imperative, therefore, that sufficient reference points be established and protected, outside the construction, to permit center-line hubs to be restored confidently in minimal time with minimal effort. Experience here is the best teacher, and no universal procedures can be spelled out.

A particularly difficult case will occur whenever a center-line point is desired somewhere along a curve. It is probable that the *P.I.* will be obscured or invisible (especially if the subgrade is down in a cut section), the *P.C.* and the *P.T.* points will possibly not be intervisible or even marked, and, unless forethought has been given to the problem, it can be a difficult one. The only general rule, thus, is to have the forethought to keep reference marks (and notes on them) so as to be quickly able subsequently to reestablish virtually any point that may be called for.

Figure 15.4.1 shows a simple set of reference stakes (*A* and *A'*, *B* and *B'*), which were preset and maintained as a means of subsequently setting *D* and *E*, two points on the center line in a deep cut. This is typically one stratagem that can be employed to replace needed hubs.

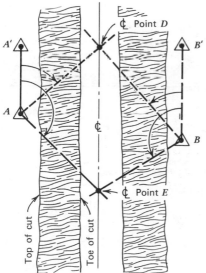

Fig. 15.4.1 One method of resetting center-line hubs from preestablished reference hubs.

15.5   Responsibility for Staking

While no specific reference has been made throughout the foregoing to the responsibility for setting and maintaining grade and alignment stakes, it is general practice that the resident engineer is obligated to set stakes initially and again finally as a check, and the contractor is obligated to maintain them and reset them during the work as needed. It is impossible to hold exactly to the rule, since conditions vary widely and unprecedented events always occur. Checking and collaboration by both parties is always desirable and necessary to assure the smooth flow of the work.

15.6   Final Staking for Finishing Grade and Paving

When rough grading has been accomplished, and the cut, fill, and other earthwork on the route is completed, new stake-out must be done to provide guidance for finishing the slopes, the subgrade, and the roadbed.

Fig. 15.6.1 Sensor on paver in contact with taut string line to govern screed height.

This is done by beginning once again from the control points. The final setting of center-line stakes to fix the profile and alignment is needed at the conclusion of subgrade preparation to serve for setting forms for pouring concrete when fixed forms are needed. This alignment could, by agreement, be a set of hubs set alongside one lane (or both), tacked in the top for line and driven to the correct level of finished grade or to exact level of subgrade ("blue tops").

The subgrade is sloped and aligned by these grade stakes and alignment stakes placed at this time at the station points and half-station points, and perhaps oftener. The paving machines will require careful guidance, especially for high-speed highways.

For paving machines laying asphaltic concrete and requiring no forms, edge-of-pavement stakes or pins that carry reference marks will guide the screed and the alignment. The same is true for portland-cement slip-form paving. A string line (mason's cord) or a taut steel wire should be used. It can be set parallel to the center line of the road, along the edge or near the longitudinal joint of the lane being paved. On tangents, the

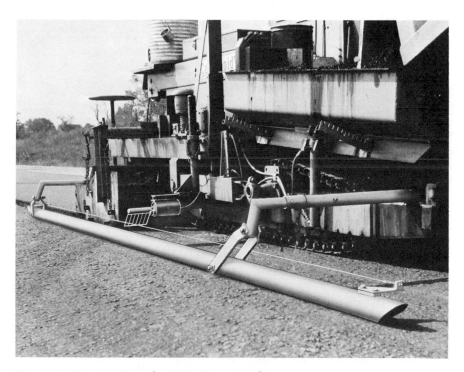

Fig. 15.6.2 Sensor guidance by "ski" riding on surface.

stakes for the line need be only at every station, and on curves, every quarter-station or every half-station. Such a line will serve to guide the form erection for concrete.

There are slip-form concrete pavers or asphalt pavers in use today that use a string or wire stretched along these stakes as a guide by means of an electronic sensor. The automatic screed control can "feel" the line and set the level, the slope, the transverse profile, and the vertical curves by following the taut line stretched as a guide. Figure 15.6.1 indicates the mechanism. The screed control operates through a grade sensor, a command box, and a pendulum designed to activate motors or cylinders to change the screed tilt, automatically compensating for roadbed surface irregularities. The sensor may use, instead of the taut line, a ski that rides along existing pavement or curb, as in Fig. 15.6.2. The string line is needed, though, for first construction.

16

Shafts and Tunnels

16.1   Verticality of Shaft

When digging a vertical shaft in earth or rock, a prime requirement is that it be kept vertical, which is somewhat difficult for a deep shaft. Using a plumb bob on a long string or wire is acceptable practice. Normally a heavy weight is placed on the end of a steel wire, and the weight is suspended into a bucket of water or viscous oil to assist in damping the pendulum sway of the wire. Whenever desired, periodically, the plumb line is set up again, and measurements are made to the sides of the shaft to ensure that it is progressing vertically. Placing reference marks, spikes, or bolts driven into the side of the shaft will enable use of these marks to carry the verticality further along by use of shorter lines for convenience.

Setting the surface mark or point is done by intersecting two transit sightings from some control points on the surface. Once found, this point is retained for future use by fixing it firmly on a bracket that can be used to support the plumb line from time to time, as will be necessary. Desirably, this point ought to be toward one side of the shaft to keep it out of the way of the work. The plumb wire is rehung periodically during the digging, certainly also at the completion of the shaft.

16.2   Use of Optical Plummet for Carrying Vertical Alignment in Construction

Transits and theodolites of modern design are equipped with optical plummets to replace plumb bobs, with the obvious advantage that they will not sway in the breeze, and so forth. To serve its function, the optical plummet must be vertical when the instrument's horizontal bubble is centered. Checking this is done by rotating the instrument through 360° about its azimuth axis while simultaneously observing that the bubble stays centered while the point observed on the ground beneath also remains centered in the field of view. The optical plummet on most instruments is good enough to center the transit or theodolite above a point some two or three stories below, provided a clear sight can be obtained. This is accomplished by building a sturdy frame to hold the tripod and instrument out over the edge of the shaft or structure. Some instruments can also be set on a special frame, or trivet, without the need of a tripod, to assist in this plumbing operation.(See Fig. 16.2.1.)

Some manufacturers can supply a special tripod-mounted vertical collimator (optical plummet) that looks either down or up vertically, especially contrived to do vertical alignment for high-rise construction or for

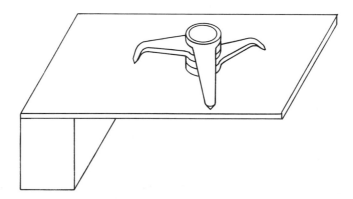

Fig. 16.2.1 A trivet mounting for a theodolite overhanging a shaft for using the optical plummet.

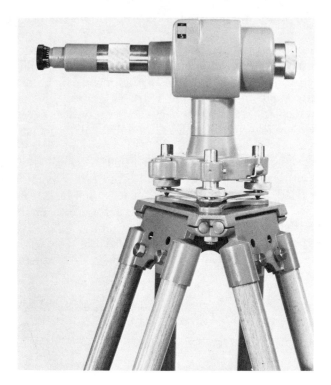

Fig. 16.2.2 Optical zenith-nadir plummet, or vertical collimator.

Courtesy Wild Heerbrugg.

sighting down deep vertical shafts. Plumbing vertically for a considerable distance requires such a precise instrument equipped with a more sensitive bubble and the capability of focusing and sighting at greater lengths. Figure 16.2.2 shows such a vertical collimator; see also Fig. 4.3.4.

By use of a vertical collinator or optical plummet, a mark can be set directly on the prepared surface of the shaft bottom, for example, onto a concrete surface troweled into readiness some short time in advance. However, measurements can be made to spots on the shaft wall in much the same manner as with the wire method; the person sighting from above can simply read the tape or rod held horizontally in position. Regular use of reference marks on shaft walls can aid the construction crew to maintain the shaft vertical by segments as they progress; a final check by vertical collimator will, of course, be necessary at the finish of the operation.

16.3   Use of Laser for Carrying Vertical Alignment in Construction

Much as an optical plummet works, the laser (see Appendix A) can be directed vertically downward or upward to project a thin light beam onto a target, as shown in Fig. 16.3.1. Alignment can be maintained by measurements from an illuminated mark at any level where the light beam is intercepted. Care must be taken, certainly, to insure that the beam is truly vertical and not out of plumb. A fixture that permits the laser ray to rotate about the vertical axis is equipped with a sensitive bubble tube. While the fixture is rotated 360° (complete circle), if the bubble remains centered and the spot on the target below (or above) remains stationary, the beam is vertical. If it is then locked into this position, the beam can be intercepted at any level for a measurement laterally to the side of the shaft or to a bracket or structure.

Washer-type targets work well, as shown in Fig. 16.3.2; these washers or plates with drilled holes are fixed in brackets to a structure or to the

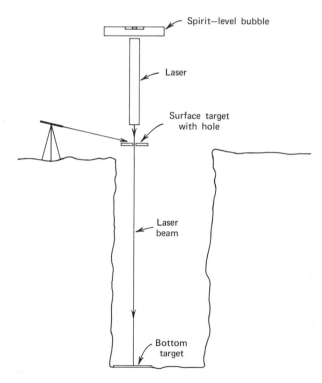

Spirit—level bubble

Laser

Surface target with hole

Laser beam

Bottom target

Fig. 16.3.1   Laser beam for dropping vertical line down a shaft.

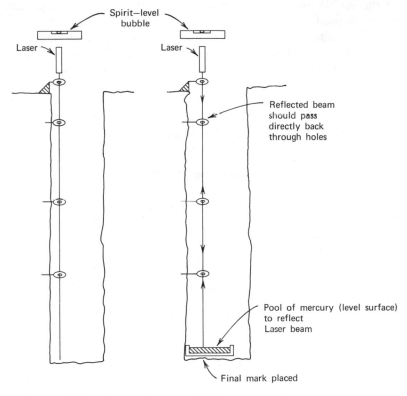

Fig. 16.3.2 Laser shining through washers fixed in shaft wall.

Fig. 16.3.3 Use of a pool of mercury as a check on verticality of laser.

side of the shaft. Using the laser ray, it is easy to center the small hole of the washer by observation. These washers on brackets then permit the laser to be quickly placed back in position if it needs to be removed, or it will be at once apparent if the laser be accidentally moved or jarred out of plumb, since the light will no longer be seen.

A simple yet positive verification that the laser beam is truly vertical can be made by placing a flat wood tray of mercury at the bottom of the shaft or structure to intercept the laser ray. If the ray is vertical, the ray of light will reflect exactly back up through the washer targets, an easily observable phenomenon. Any slight error in verticality of the beam will be immediately evident and can be corrected. The mercury reflects the ray very well, and is a truly level surface that rapidly becomes still. Once used, the dish of mercury is removed and a mark can be placed in the center of the visible ray on any prepared surface below. This procedure could likely be used as a simple and feasible method of transferring a mark from the surface to the bottom of a shaft, although some details of laser beam width, procedure, and so on, may still need to be refined in each instance. Different types of lasers having narrower beams and greater intensity must be used for deep shafts, for example, of 600-ft depth, since the light must travel double that distance in this method. Greater experience will be needed with this new procedure, although it is basically sound. It is sketched in Fig. 16.3.3.

16.4   Tunnels and Shafts

Tunnel alignment is a particular specialty, requiring techniques of surface surveying and mine surveying, plus considerable improvisation and knowledge of tunneling procedures. Essentially, horizontal alignment must be dropped (usually) in a vertical shaft of limited diameter. The objective is to link the underground survey with the control survey at the surface so that the tunnel bore will properly follow the planned alignment. Usually two points are dropped (A and B in Fig. 16.4.1) that serve to connect the surface survey with the tunnel survey. From the two points, underground alignment is carried forward in the tunnel as the boring, drilling, and mucking progress. There is usually very little opportunity for independent checking to the surface by means of other shafts or access tunnels, so great care must be employed in dropping line. Once in the tunnel, alignment hubs are installed along the tunnel sides or roof for accessibility and security, while on the surface the control survey is carried forward by regular methods. When the tunnel correctly reaches another shaft, of course, or when two segments of the tunnel meet, this is the first check of the tunnel alignment and the first opportunity for any minor adjustments. Fig. 16.4.2 shows an underground alignment survey in progress.

16.5   Dropping Line in a Shaft by Plumb Bobs

Dropping a horizontal line from the surface to the bottom of a shaft can be done with two piano-wire plumb lines, heavily weighted and stilled in buckets of water or oil, plus patient multiple observations during a lull in operations when no ventilation drafts affect the plumb lines. Figure 16.5.1 gives some indication of the elaborate procedure. The supporting structure for the plumb wires must provide for minute lateral movement of each wire while being aligned by the transit or theodolite at the sur-

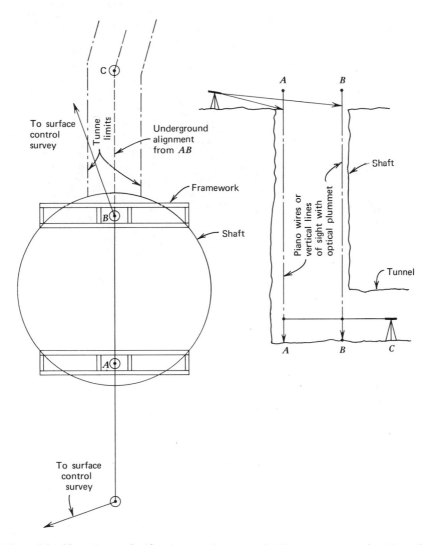

Fig. 16.4.1 Top view and side view to show tunnel alignment connected with surface control survey; alignment dropped to tunnel in vertical circular shaft (the transit at C observes the wires at A and B and is itself shifted into correct alignment).

Fig. 16.4.2 Tunnel alignment carried forward by transit on tripod.

Courtesy Kern Instruments.

face. Below, the transit again observes the wires, against a graduated target background, to set hubs for aligning the tunnel.

Considerable care in placing plumb wires and weights must be employed to assure that the wires are not fouled or displaced in the slightest by any projection of the shaft, brackets, pipe flanges, timbering, and so forth. They must be protected from spray or spatter of water and from leaks from air pressure lines or ventilating pipes. During line-dropping it is necessary to stop the elevator operation and probably turn off all ventilation blowers. It is a critical operation calling for sole and exclusive use of the shaft to avoid all disturbances. Standard procedures for this work are described in specialized manuals developed by organizations regularly doing such work.

16.6   Dropping Line in a Shaft by Optical Plummets

Using two optical plummets mounted above and sighting down the shaft to two adjustable targets can also serve to do the job of dropping an

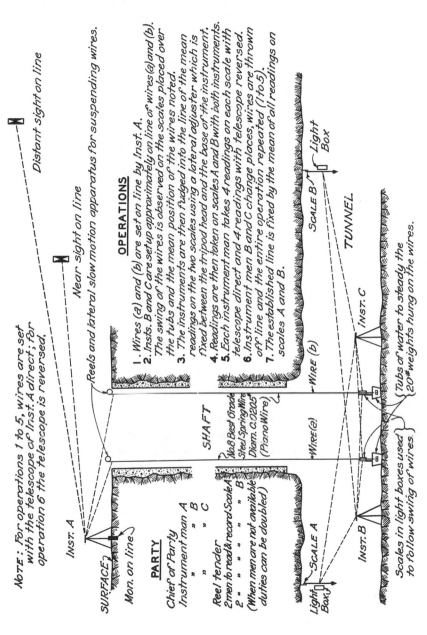

Fig. 16.5.1 Tunnel alignment line dropping procedure by N.Y.C. Board of Water Supply.

alignment for the tunnel. Experience with this method is limited as yet. Adapting standard instruments for this sort of operation should assure adequate accuracy; some improvising of accessories, usual on any such task, ought not to be too difficult.

The optical plummet of Fig. 16.2.2 (shown in cross section in Fig. 4.3.4) is capable of great accuracy, and several manufacturers supply such equipment. Plumbing accuracy of such an instrument carefully adjusted and carefully handled is in the order of 1:30,000 or so. The procedures involved can rather quickly be adapted to American practice; experience with shafts of varying depths will soon adduce operational standards.

16.7   Dropping Line in a Shaft by Telescope with Right-Angle Prism

Another new device used for "plumbing" shafts, the automatic level equipped with a rotatable right-angle prism at its objective end, may also prove its worth for dropping two points for aligning a tunnel. (See Section 4.10.) No experience with shafts is reported as yet, but the principle is sound, having been used for similar work in Europe. One places the level, centered over an approximately correct point at the foot of the shaft and observes upward to a specially prepared graduated cross. By taking multiple readings of the four target scales (one on each arm of the cross), a mean point can be discovered, which is exactly above the center of the instrument axis. See Fig. 16.7.1, although details may still have to be worked out.

Repeating the same process for another selected point at the foot of the shaft gives a second point on top. The alignment on the surface can be ascertained by tying the two points to existing control at top; any calculated variation from true alignment, which should be slight, can then be adjusted below ground by the theodolite in use. The principle is shown in Fig. 16.7.1, with the instrument at the bottom of the shaft. It would seem more logical, in fact, to use this method with the automatic level and prism set on a firm mounting at the surface and sight to target scales below.

In either case, the important aspect of this method is that the compensating mechanism of the automatic level will hold a level line during each sighting, reflected in good vertical alignment while sighting in four directions below (or above). Thus, along with successive multiple readings, the results should be good.

Transferring alignment from the surface down a vertical shaft can also be done by a theodolite fitted with a right-angle prism (Fig. 16.7.2) at

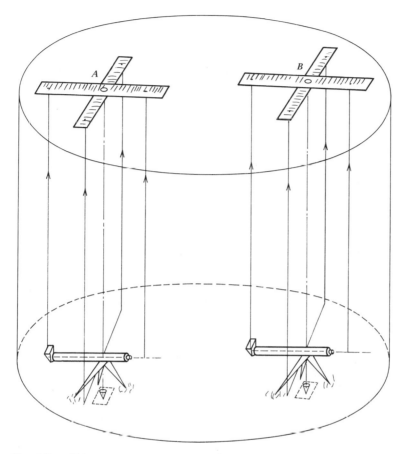

Fig. 16.7.1 Using automatic level with 90° prism to drop line in a shaft.

the objective end, which rotates about the axis of the telescope. In Fig. 16.7.3, point A, on the line between P and Q, is set at the surface on a securely mounted frame that will not be affected by the instrumentman's movements. At the bottom of the shaft two finely divided scales S_1 and S_2 are placed at right angles to the line being transferred, and as far apart as possible.

The theodolite is then set carefully over A with telescope horizontal, sighted on point P through the 90° prism, and then clamped. Then the prism is rotated to point downward, and the scales S_1 and S_2 are read. Without changing the inclination of the telescope, the prism is again turned horizontal, facing opposite to its first position. The theodolite is

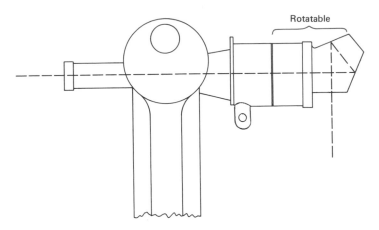

Rotatable

Fig. 16.7.2 Schematic of right-angle prism mounted on theodolite objective.

rotated about its vertical axis until point P is again sighted, and the horizontal motion of the theodolite is clamped. Once more the 90° prism is rotated so it looks down to the scales, and readings on S_1 and S_2 are again taken.

Now the point on S_1 which is the mean of the two readings on that scale, and the point on S_2 which is the mean of the two readings on that scale, will form a line that is exactly parallel with line PAQ above. An instrument below ground (in the tunnel) could then be aligned with the correct midpoint readings on the two scales and the line would be effectively transferred down the shaft. However, prudence would dictate that the procedure be repeated, this time with the telescope in inverted position as a verification. In fact, because of the importance of such an operation, the entire sequence might well be repeated several times to get the mean of many sets of readings.

The principle involved is, of course, that the 90° prism attached to the theodolite is carefully ground and fitted so that it is a true right angle, and the line of sight sweeps a true plane from point P down to the scales. Interestingly, this procedure calls for only one instrument setup at the top of the shaft instead of two. Results by this method depend upon the care employed and the inherent accuracy of the instrument. Details of this relatively new procedure would have to be worked out to account for the many variables (depth of shaft, requisite illumination of scales, means of achieving clear sights, etc.), but there is no reason to expect that it cannot give very satisfactory results.

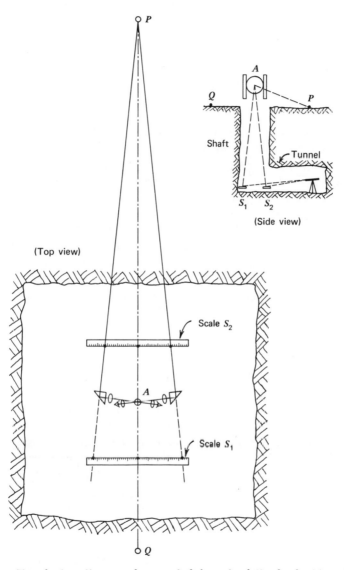

Fig. 16.7.3 Transferring alignment down a shaft by a theodolite fitted with a right-angle prism.

16.8   Dropping Line in a Shaft by Lasers

Because using laser beams to fix two points at the bottom of a shaft to carry alignment has not yet been established by enough practice, not much can be said about the results. In theory, the use of the pool of mercury at the bottom of two vertical lasers should establish a near-perfect aligning, but there are practical considerations such as narrowness of beam width, the means of transferring the laser beam to an accurate point, and the aligning of a transit or theodolite within the tunnel. Only time will tell, but the method holds promise.

Some amount of bending can occur in the laser beam by refraction caused by layers of air of different temperature, and there can be some scintillation visible on the laser target that seems to be traceable to turbulent air. Experience is too limited as yet to fix firm rules about the phenomenon, but it would seem that in plumbing shafts it may be well to keep the ventilation system running and avoid having the rays in close proximity to warm pipes. The phenomenon is of course identical to that induced when light rays shimmer, nothing more mysterious, and would occur in the use of optical sighting methods described in Sections 16.6 and 16.7 as well. There is nothing inherently different in the laser that would bar its use in shaft work therefore.

16.9   Establishing Vertical Control in the Tunnel

To assure correct elevations for shaft and tunnel, vertical distance is best measured down the shaft with a one-piece steel tape of sufficient length. Hanging a 20-lb weight on it will assure that it will give correct length, although the tape will have to be calibrated at proper tension and any temperature correction will have to be applied. Interestingly, if the tension at the bottom is 20 lb due to the weight, the tension at the top of the tape will be 20 lb plus the weight of the tape. Calibration should be done at the average of top and bottom tensions, therefore, when comparing the working tape with the standard tape.

Bench marks, more than one, must be available near the top of the shaft, and new ones must be prepared at or near the foot of each shaft. On signal, simultaneous readings of the tape are taken by engineers' levels at the top and at the bottom of the shaft, as seen in Fig. 16.9.1. Several sets of readings are made, with the tape raised or lowered between each set to avoid prejudiced readings by the instrumentmen. The mean of the several sets will give the difference of elevation between

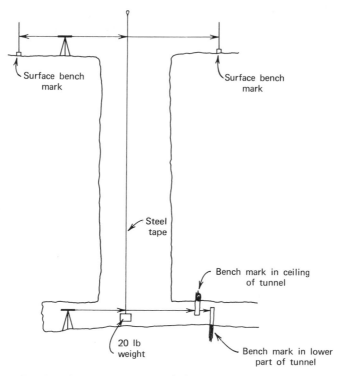

Fig. 16.9.1 Carrying elevations down a shaft for vertical control in a tunnel.

the telescope lines of sight, from which calculations to bench marks can be made.

As the tunnel is carried forward tunnel bench marks are placed, usually metal spikes or bolts suitably headed for holding the rod, firmly cemented into holes drilled into the rock at top, sides, or bottom of the tunnel. As new shafts are driven, check measurements must be made by leveling forward on the surface and dropping a tape down the new shaft to check the tunnel bench marks that have been carried forward. Any elevation adjustments can then be made, much as any level circuit would be adjusted (see Section 2.13).

Bench marks set in the roof of the tunnel have been found to move downward as the rock moves to relieve stresses. Consequently, more reliance should be placed on bench marks set in the sides of the tunnel, or on special concrete pedestals constructed along the bottom and side of the tunnel. At all events, when carrying elevations forward, two or more bench marks should always be used for security, since bench marks in a

tunnel are always subject to accidental disturbance anyway. Bench marks
are needed every 200 or 300 ft as a rule to control the tunnel excavation
and finishing.

16.10   Alignment in Tunneling

The principal need to carry alignment by surveying methods within the
tunnel is to assure that the direction and slope of the tunneling will be
correct. Consequently, much reference must be made to line and grade
points; frequently one must mark alignment up close to the heading, as
also on the tunneling machine itself to keep the bore straight and true.
Furnishing proper grade or elevation information for setting tunnel liner,
concrete work, or finish floor is equally important and demanding. Because neither line nor grade can be well checked until an adit or another
shaft is reached, or until "holing through" occurs, great reliance must be
placed on measurements tied in at only one shaft or adit. After a next
checkpoint is reached, however, things get somewhat better, inasmuch as
there is a possibility of back adjustments, if needed.

Tunnel surveys require external control above ground of a high order
of precision and the transfer of line and grade down shafts with great
care. It is most helpful if alignment can be introduced into the tunnel by
a horizontal or nearly horizontal shaft or adit, as in Fig. 16.10.1, simply
to avoid the always uncertain dropping of alignment in a small-diameter
shaft.

Line is carried forward in tunneling by tape and theodolite to marks
placed either on the floor or side or roof of the tunnel. Driving plugs
and spads into the roof to serve as horizontal control (traverse) points
keeps them free from traffic damage or disturbance, as in Fig. 16.10.2.

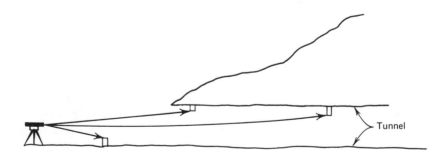

Fig. 16.10.1 The simpler case of carrying alignment into a horizontal tunnel.

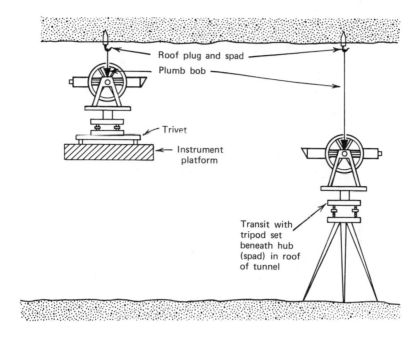

Roof plug and spad

Plumb bob

Trivet

Instrument
platform

Transit with
tripod set
beneath hub
(spad) in roof
of tunnel

Fig. 16.10.2 Use of roof hubs in tunnel to avoid disturbance by construction activity.

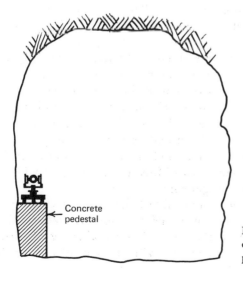

Concrete
pedestal

Fig. 16.10.3 Tunnel alignment carried on concrete pedestals, with transit placed on trivet above the control hub.

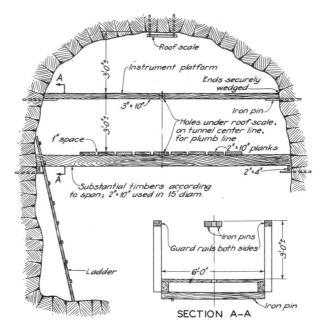

Fig. 16.10.4 Typical tunnel alignment instrument platform in unsupported tunnel.

Some tunnels are wide enough to permit an alignment offset to one side, as in Fig. 16.10.3. Sometimes an elevated platform is built to raise the transit to a semipermanent position above the flow of traffic, as in Fig. 16.10.4, or a pedestal is constructed on one side of the tunnel for fairly permanent control.

Carrying line forward in a tunnel will normally require the use of scales mounted in the roof of the tunnel, or at each hub some device must be provided to permit lateral movement of the point. Precision targeting (and illumination) will be required. The scales permit prolonging a straight line by double-centering (as in Section 4.8) with direct and reverse readings on the scale serving to provide a mean reading, which then is the true alignment. Driving a hub and spad for a plumb bob is not really precise enough for carrying control forward, although it will be usually sufficient for the contractor's guidance.

Control points (scales or laterally adjustable hubs) are set every 300 to 600 ft, depending on visibility in the tunnel atmosphere, whereas the contractor may well need guide points every 75 ft for aligning his drill rig and/or shield at the face of the heading. With the advent of tunneling machine (moles) today, however, a more precise control may have to be made readily available at any instant to guide the machine.

In case a tripod cannot be used, a trivet or base plate attached to the instrument can provide a means of mounting it on the shelf platform or even the overhead structure. In the event that a ceiling hub (spad) is used, the transit is centered under a suspended plumb bob. See Fig. 16.10.2 for two variations thereof. Transits and theodolites used in underground work require a center mark or dimple on top of the instrument for centering beneath a plumb bob. When this mark is used, the centering of the instrument is true only when the instrument is in adjustment and the telescope is level.

The responsibility for alignment and grade is very great in a tunneling operation because of the great cost of an error. Investing extra time and money for extra precision is always warranted. The use of a three-dimensional coordinate system can also be of great help in maintaining control of the job and furnishing information whenever it is needed.

A new method of orienting a line in a tunnel is possible today with the perfection of gyroscopic theodolites. While expensive, their use can save expensive efforts to orient a tunnel and maintain correct alignment, or to check alignment. The principle is that a spinning gyroscope is suspended with its spin axis kept in the horizontal plane, and the earth's rotation will cause a directional moment turning the spin axis toward the plane of the meridian (the north-south plane). Should the inertia of the mass attached to the system cause the spin axis to pass through the meridian and go to the other side, there will be a directional moment in the opposite direction turning it back towards the meridian. (See Fig. 16.10.5.) This precession movement is a sort of oscillation, and at each reversal of the oscillation, where an apparent standstill is observed, a reading is taken on the horizontal circle of the theodolite.

By computing the mean of a succession of consecutive reversal point circle readings, a circle indication for true north is obtained. From the true north reading, a bearing for the tunnel heading can be calculated and set on target points or scales. If a gyro-theodolite is used in a tunneling project to establish alignment, naturally only one point need be dropped down the shaft, not two. The gyro can ascertain the direction for the tunnel bore by furnishing a true north-south meridian when set up on the single point dropped down the shaft. Multiple observations at this point, as time permits, will refine and/or verify the initial direction determined.

It should be noted that the direction furnished by the gyroscopic instrument is a true azimuth, or bearing, based on the earth's spin. In the

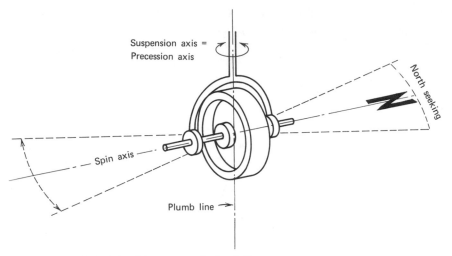

Fig. 16.10.5 Principle of the gyroscopic theodolite.

event that the grid system being used in the construction is based on some arbitrary north-south meridian, a correction will have to be applied to the gyro directions down below. Probably one or more observation with the gyro-theodolite on a control station at the top of the shaft will best be made to measure the angular difference between the true north and the grid north being used. If the underground project extends any considerable distance east or west, the true meridian and grid meridian can differ noticeably, since the true meridians tend to converge while the grid meridians are by definition parallel. This is illustrated in Fig. 16.10.6.

The formula for convergency of meridians is:

$$C'' = 52.13\ L \tan \phi$$

where

$$C'' = \text{convergence (sec of arc)}$$

$$L = \text{east-west distance (mi)}$$

$$\phi = \text{latitude of the place}$$

Since the gyroscopic bearing can be obtained to within a few seconds, this convergency of meridians can become important. If at latitude N45° a tunnel extends 1 mi in the east-west direction, the discrepancy would be 52″ of arc, enough to cause some real problems unless anticipated. A

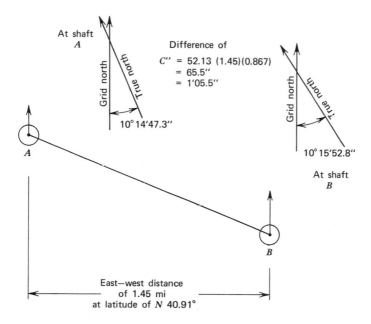

Fig. 16.10.6 Convergency of meridians in tunnel alignment using gyroscopic theodolite methods.

sample of an actual tunnel project in the latitude of New York is shown in Fig. 16.10.6 in which the meridians converge sufficiently to give cause for concern.

By reversing the reasoning, one can see the need to study out the procedure of adjusting bearings or azimuths underground if one were to use gyroscopic theodolites from two shafts intending to have two tunnel headings meet halfway. The problem is readily solved by being aware of it and simply prolonging grid bearings and using grid coordinates underground by careful methods.

The north-seeking gyro is adaptable to various types of theodolites and tripod setups, or can be adapted. Tunnel aligning by use of the gyroscopic theodolite is recommended on projects of considerable cost and where very little opportunity exists for frequent checking at shafts or adits. Checking by the gyro will tend to reduce cumulative errors, which can creep in with short backsights or foresights—such as will happen in curved tunnels, for example.

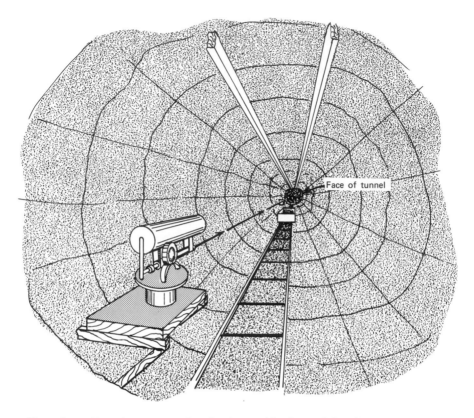

Fig. 16.11.1 Laser beam mounted on bracket at side of tunnel for alignment.

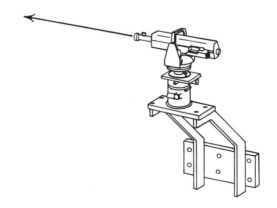

Fig. 16.11.2 Laser with elevation and shifting heads attached to rigid bracket.

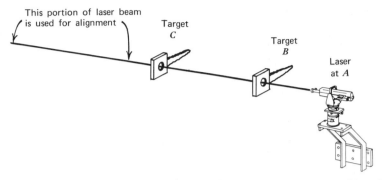

Fig. 16.11.3 Using extra targets (holes) for easy detection of misalignment of laser beam.

16.11 Use of Laser Beams to Guide Tunneling

As a means of aligning equipment and furnishing grade and line to all and sundry who constantly need such information, the laser beam has come into use quite satisfactorily. It does not substitute for the surveying (angle-measuring, taping, differential leveling, etc.) in the tunnel, or the placing of control bench marks and spads in the roof for alignment traverses, and so forth, but it has a practical function in providing immediately accessible information. By setting a laser beam to point in a direction parallel to the tunnel axis and at the proper gradient, one can supply line and grade easily. The beam need not be in the center of the tunnel, and generally would be in the way if it were. It must be made parallel to the axis, though, as shown in Fig. 16.11.1.

The laser is generally mounted 2 to 3 ft off the wall by a bracket attached to the liner plate, or bolted directly into the ribs, or into the rock. (See Fig. 16.11.2.) Once in position, the beam is projected to the face. Two points between the laser and the face define line and grade; a target may be placed at one or more points, as may be seen in Fig. 16.11.3. The target is made of steel or wood, circular or rectangular, with a $\frac{1}{2}$- to $\frac{3}{4}$-in. hole drilled directly in the center. When the target is properly adjusted and the beam passes through the hole in the target, the laser beam then parallels the center line of the tunnel and becomes a visible line and grade reference. It can be intercepted by a person who wishes to measure distance sideways in the bore to find the center line, or who wishes to measure up or down therefrom to check his elevation. It does not require an operator to sight through a transit or a level and shout or wave, a great boon in the loudness of the tunnel. Being itself illuminated, the laser eliminates an auxiliary flashlight or torch to make a reading. And any slight interruption of the ray of light will not really inconvenience someone farther along who may be wishing to use the beam at nearly the same time. As the heading progresses, the laser is moved ahead so as to be never too far behind the working face. Lasers have worked tunnel alignment as far as 1600 ft ahead, but advancing the laser instrument forward each day is better, to avoid the deflection of the beam by reason of temperature differentials in the layers of air in the tunnel.

Where a hydraulic shield, mole, or other rock boring equipment is used, lasers can guide the machine with a continuous visible line and grade reference, for the laser beam is set to show on a guideboard facing the operator of the mole or drilling rig. This is shown schematically in Fig. 16.11.4, where two targets are mounted on the excavating machine to insure it will not become angled in the heading. The rear machine target has a hole drilled directly in its center to allow the beam to pass through. With a set of cross hairs scribed on the forward target, light from the laser strikes the intersection of these cross hairs; when the beam references exactly, the machine is exactly on line and grade.

The operator simply watches the spot of light. When it shows on his target, he can ascertain that he is on correct line and at correct grade, or can make any needed adjustments to bring his machine back if it be straying. If, for example, it moves off to the upper right, the operator knows the boring machine is being pushed to the left and down; in addition, he knows the approximate amount it is out of alignment. Absence of the light will alert the machine operator to trouble, such as will occur when the machine (and the holes) get far out of line, or if the light source is accidentally disturbed.

Problems arise when the drilling must progress around a curve—as may well be the case with either water tunnels or transportation tunnels—inasmuch as the drill rig must maintain a curved alignment. The laser can handle this, though it requires more frequent changing of the alignment holes. Where spirals, vertical curves, or horizontal curves are encountered, there are some differences in the positioning of the laser and in the type of targets used, but essentially the process is as outlined for tangent headings. Since the bulk of tunnel work is straight-line drilling, with lesser amounts of curve work, this works fairly well to have the laser as a guide.

Going one step further in using the laser, machines are being developed that can be guided automatically by the laser beam. Equipped with sensors to catch any straying from line or grade, they operate relays to actuate hydraulic jacks or electric motors to reset the alignment. The

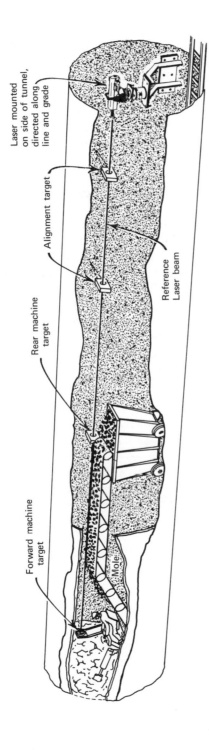

Laser mounted on side of tunnel, directed along line and grade

Alignment target

Rear machine target

Reference Laser beam

Forward machine target

Mole

Fig. 16.11.4 Laser beam used as a guide for tunneling machine by use of two targets mounted on the machine.

229

laser beam can be easily caught by a mirror and reflected back from the front of the tunneling machine onto a target, even a translucent plexiglas target for easy viewing; sensors mounted on this target board can be placed to guide the mole by automatic relays—or the operator can, if convenient enough, rely on his manual control, anyway.

Experience with laser control in mining and tunneling indicates that cost reduction is realized in total labor requirement. Considerable savings are realized also in the amount of time the engineer spends on the project. Improved accuracy and reduction or elimination of human error by survey parties or operating personnel in the heading will result in dollars saved.

When setting up a laser for alignment purposes, one might be aware of a need to use three, not two, points to form a straight line. For example, if in Fig. 16.11.3 there be only the target washer at B and the laser position at A to form the straight line, should the laser generator be moved slightly and then realigned so its beam passes through target B, there would be no assurance of its correct alignment. If, on the other hand, the alignment were to be fixed at the start by the laser position A, and by two other targets or alignment holes at B and at C, then misalignment would be noticed at once. Any bumping or jarring of the machine at A could be detected and corrected by resetting the laser to shoot its beam through the target washers at B and C.

17

Offshore Measurements

17.1 Locating a Point Offshore

Locating or fixing an offshore point, a not uncommon need, can be done in a variety of ways. The intersection of two targeted range lines can be seen from a boat; two distances can be simultaneously observed on a boat equipped with a Decca-system navigating device or with a Hydro-Dist installation; two transits, oriented on shore-based control, can be pointed to sea for coordinated observation of a boat or target buoy; simultaneous or nearly simultaneous sextant observations from the boat can be made of the two horizontal angles between three known shore points for locating the boat; and in some situations, lasers can do the job.

The need for locating oneself offshore arises when making soundings (Fig. 17.1.1), when dredging, when placing (or finding) an underwater structure, when setting bridge piers, drill rigs, or pile clusters, or when placing a temporary structure for a surveying platform to control construction layout. Placing such an important thing as a bridge pier will typically call for a temporary platform from which measurements could be made during erection. As the work on the structure progresses, fre-

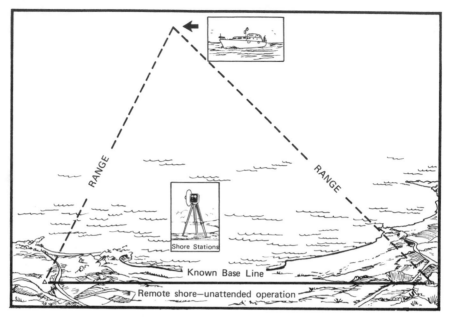

Fig. 17.1.1 Locating a sounding boat's position with EDM equipment that utilizes radio frequencies (master station on boat, with slave stations ashore).

quent rechecking and refinement of measurements from the platform and from shore would normally be done.

17.2   Methods to Fix Offshore Points

The method of getting ties to shore control or base lines is selected by consideration of such factors as distances involved, accuracies needed, and equipment available. EDM equipment that uses radio waves can be satisfactorily employed on a moving boat, whereas EDM equipment that uses light waves cannot, because of the rocking motion of a boat. On the other hand, either type can be employed on shore to locate fixed points, such as carefully anchored boats in calm water, survey platforms, and drilling platforms. There is another little problem in using radar-frequency EDM; such waves sometimes exhibit peculiarities or anomolies when traveling close to a water surface, whereas light waves do not. It is easy enough to overcome these difficulties, however, and virtually all surveying for piers or piles or drilling rigs at any appreciable distance from shore is done by EDM.

Lasers, besides being the modulated carrier basis for some electronic distance-measuring equipment, are increasingly used for alignment of bridge piers over water, replacing the transit or theodolite to an extent. In the usual case, a permanent base is built for the laser on one shore and a fixed target on the far shore. Whenever alignment is needed at any pier of the bridge, any appropriate target device is moved into the path of the laser and lateral measurements are made for setting forms, bolts, steelwork, and so forth.

It should be noted, however, that a fixed laser beam aimed from shore is of no use to keep a boat moving along a straight line while making a fathometer survey or taking soundings. Those on the boat cannot see the beam except if they are exactly on line, and so they cannot tell whether to move right or left to find the beam.

17.3 Angle-Positioning Equipment

A new device just coming onto the market, however, may well function for aligning a boat making soundings along a given straight course. This "angle positioning equipment" is an electro-optical device that emits a beam of intense flashing light visible over a considerable distance in bright sunlight. (See Fig. 17.3.1.) The beam consists of three parts: a narrow, fan-shaped segment in the center, and wide steering segments on either side of the center. Each side segment has its own observable flash code sequence to guide the observer to steer right or left in order to get back on line. It is not a laser, though laser beams could conceivably be thus mechanically adapted to perform the same function.

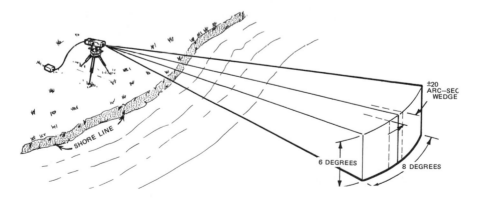

Fig. 17.3.1 Angle-positioning device using coded light flashes.

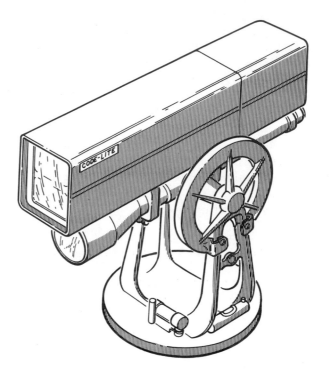

Fig. 17.3.2 Angle-positioning equipment mounted on a transit for alignment.

Courtesy Sanders Associates, Inc.

Mounted on a transit that is aligned correctly, as in Fig. 17.3.2, this positioning device uses a flashing xenon light, projected optically into a 6 by 8° rectangle from atop a tripod (or from atop a transit). In the left-hand sector it encodes a two-dot code, in the right-hand sector a one-dot code, and in the center 20 sec of arc, a combination of all the dots. Hence, by looking at the flashes of light, one knows which way to go to approach the center, to get on line. Locating and maintaining position within the beam is simple because there is a visual reference that can be detected at the point where the positioning information is needed.

In any application where angular position or direction is presently maintained by visual sighting through a transit or by a ranging line, this offers an alternative that should ease the problem. Where a level (or even an inclined) plane is now maintained by staking from an initial elevation, by use of level sighting, and so forth, the instrument suggests itself as a substitute by being rotated sideways 90° and flashing its lights.

17.4 Cuts to Range Lines

Nevertheless, simple stratagems for locating a boat's position can be contrived. Ranging poles can be erected in pairs to mark a line. As shown in Fig. 17.4.1, for work within a mile of shore, a sounding boat can steer a course fixed by range targets on land. Farther from shore the boat can best be kept straight by having a transit set on the range line and having the instrumentman talk to the boat by radio as it travels. This operation is shown in Fig. 17.4.2. Another one or more transits located some distance off the range line can follow the boat as it moves and makes simultaneous "cuts" on its position as it is periodically cued by a color-coded signal flag from the boat or by radio. The color coding or some similar identification is almost a must in order later confidently to correlate the notes taken by widely separated notekeepers, one at each transit recording angles and time, and one on the boat recording sounding depth and time.

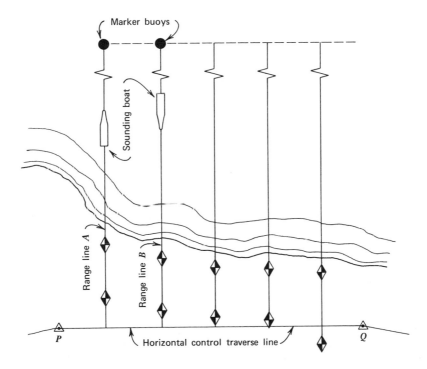

Fig. 17.4.1 Sounding boat following range lines fixed on shore while moving at a fixed speed, making soundings at fixed time intervals. Buoys are an added convenience, and can be located from shore control points *P* and *Q* by triangulation.

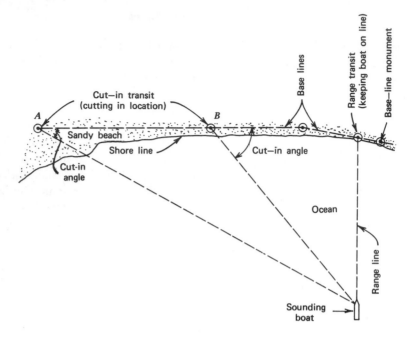

Fig. 17.4.2 Sounding boat being guided along range line by transit on shore using hand signals or radio instructions. Additional transits make simultaneous "cuts" to boat on signal, to fix boat's position at given times.

On a clear day an instrumentman on a 50-ft shore tower (or hill) can see a boat nearly 10 mi out from land. Simultaneous "cuts" from shore points, traverse stations, or triangulation hubs can be made to locate such a moving boat easily within 10 or 20 ft. The two or more instrumentmen would best be cued by the radio signal for these great distances, and the instrument stations should be far enough apart to afford strong intersections of sight lines for later plotting.

17.5 Dredging by Range Lines

Dredging by conventional control (without lasers) is accomplished by providing on-shore sight lines (range lines) by which the dredge master aligns himself. Each range line consists of two matched (coded or numbered) targets of sufficient size to be seen from the dredge. These are set by transit survey. Intersecting range lines mark the beginning or end of channel segments, limits of dredging area, and so forth, as in Fig. 17.5.1.

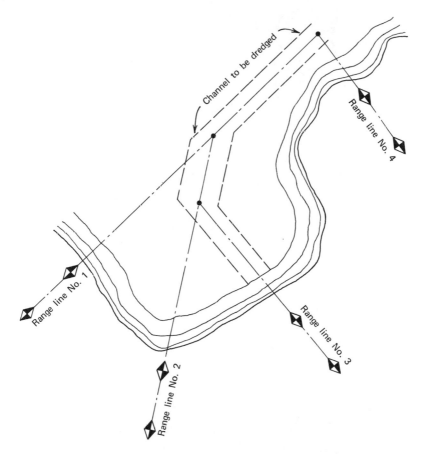

Fig. 17.5.1 Range line intersections to guide dredging by self-alignment from the dredge vessel.

Depth of dredging is controlled from the water surface, in tidal areas being correlated with tide-gauge readings on the shore.

17.6 Dredging by Lasers

In trenching or dredging beneath a water surface to virtually any depth or by virtually any machine, the laser is also useful for controlling alignment and depth. The range lines of Fig. 17.5.1 could be laser lines, for example. The dredge operator can align himself with a laser shining from shore because his boat is virtually stationary. If, in addition, another laser

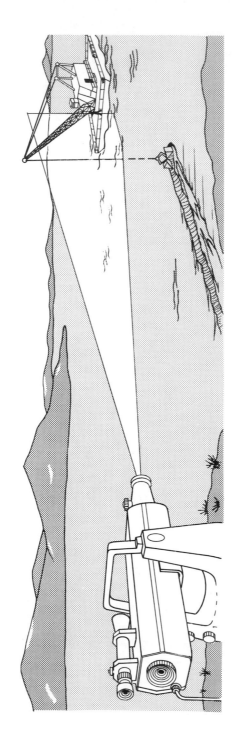

Fig. 17.6.1 Dredging for a submerged outfall pipeline by using a laser equipped with a fan lens mounted to give a vertical line.

beam (a fan laser) is sweeping the horizon to give elevation control, the dredging becomes independent of tidal action. A narrow-angle fan lens can be horizontally mounted in front of the laser instrument to furnish a horizontal plane at a given elevation, and the dredge operator will see the light intercepting a vertical rod affixed to the dredge. He can thus control his dredging depth. Refer also to Section 17.7.

Such lasers can control dredging for a considerable distance from shore in a harbor or estuary. A laser beam is used, for example, to accurately dredge a trench into which an underwater outfall sewer pipe or a prefabricated tunnel will be laid. Figure 17.6.1 shows a trench-alignment laser with an accessory mounted to give a vertically fanned beam to keep the dredge on line. The fan lens increases the height of the beam, but not its width, making the correct path easy to locate and follow regardless of tides and wave action. When a prefabricated tunnel or pipe section is lowered into place into the trench beneath the water, survey towers mounted temporarily atop the section can be used as targets to verify alignment by either transit or laser observation from shore as it is being settled to the bottom.

A large-scale use of the laser for offshore dredging of a trench is illustrated in Fig. 17.6.2, as worked up by Richard Millard of Trans-Bay Constructors (a joint venture) for use on the Bay Area Rapid Transit (BART) tunnel across San Francisco Bay. A laser beam was fanned vertically to strike the dredge on the handrail of the catwalk, and another was fanned horizontally to strike the vertical ladder. Both ladder and handrail were marked off for distance so that both horizontal and vertical control were available for positioning the dredge and digging to the correct depth. The dredge operator translated these laser beam readings, which were his only control for the 5-mi distance.

An interesting self-checking method was used in this case: a retroreflector placed in line on the opposite shore. The dredge operator could at any instant, day or night, verify that the laser beam was still properly aligned if he could see the laser reflected back along the line. As an alternative, had there been no convenient place for a retroreflector on an opposite shore, the beam could have been aligned through a hole bored through an opaque target—or through a narrow slit in one or more targets near the laser instrument. In any such situation, it should be apparent to the dredge operator when the laser has been disturbed and is not giving correct alignment. This safety measure will prevent the dredging from proceeding erroneously on a misaligned beam.

It will be noticed from Fig. 17.6.2 that the vertical laser beam in this case had a horizontal spread of 4 in. per mile, but this is not a serious

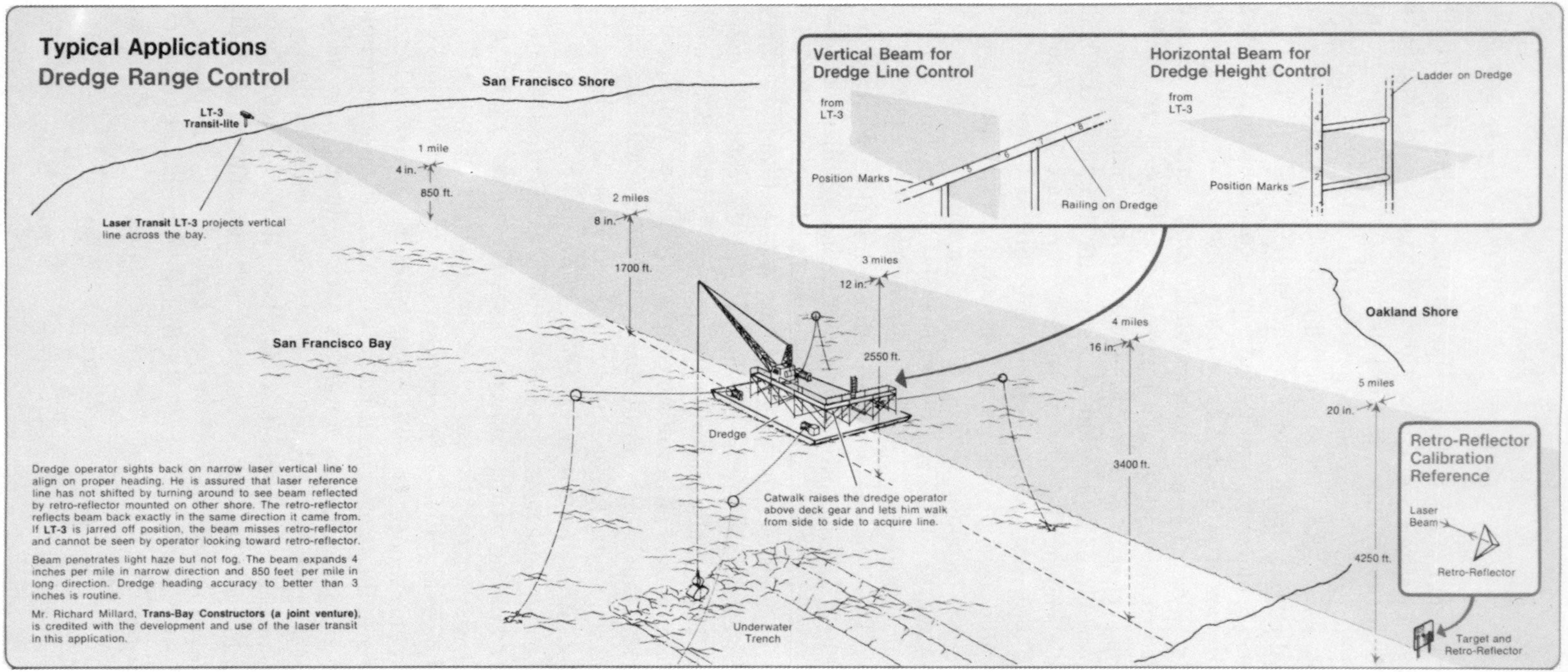

Fig. 17.6.2 Laser control of offshore dredging.

240

problem for the purpose intended. It means, however, that during the daytime there would be insufficient laser beam intensity for the red light to show on the catwalk railing. Instead, the operator could walk along the catwalk until he observed the beam by eye and mark the handrail.

The other laser beam, fanning horizontally to control depth of dredging, intercepted the vertical rail of the ladder and was observed in a similar manner. However, because of the distance involved, this horizontal light beam is subject to curvature and refraction corrections as the earth curves and the beam is bent or refracted by the atmosphere. The formula for this correction is the same as for use in using the engineer's level for long sights:

$$C_{CR} \text{ (ft)} = 0.574 \, k^2$$

where

$$k = \text{distance (mi)}$$

This is the same formula as given in Section 2.1; the laser beam follows the same refraction law as do ordinary light rays.

17.7 *Horizontal-Sweep Lasers*

A laser can sweep a horizontal plane by means of a rotating angled mirror. Full horizontal fanning or sweeping is accomplished by having the laser beam mounted vertically in a housing equipped with a rotating right-angle mirror or prism, much like the familiar rotating beacon mechanism on a police car. The rotation or scan angle can be made to vary from 0 to 360° as needed. Set on shore and carefully monitored, the beam can guide dredging depths, especially in tidal waters. Other applications suggest themselves, although not all have been tried yet: as a guide for cutting off piles to a uniform elevation on a large job; as an aid to grading an airfield or a large playing field, in which case the sweep plane could be inclined to provide for the natural runoff slope desired; as a rapid means of setting marks on columns in a building as each new floor is added; and others.

Figure 17.7.1 shows a laser beam in use to furnish horizontal alignment for driving a long line of piles offshore. It is possible that a lens to fan the beam vertically could assist the pile-rig operator in placing and driving the piles, and subsequently the lens could be turned 90° to fan the beam in the horizontal plane for marking the piles for cutoff. However, as explained in Section 17.6, care must be exercised to apply corrections to elevations by reason of curvature and refraction if the horizontal distance becomes very great.

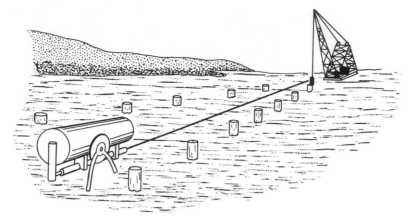

Fig. 17.7.1 A transit-mounted laser used to drive a long row of piles.

18

Random Field and Office Techniques

18.1 General

A number of methods, devices, or procedures can be contrived that may at one time or another prove helpful in construction measurement. Some are mentioned or briefly described here as a potpourri of leftovers.

18.2 String Lining of a Field

Grading of a playing field, tennis court, parking lot, lawn area, or even the subgrade for a building can be aided by sighting of strings stretched along on opposite sides of the work, with a preset T-shaped sighting board carried about and placed on the field for checking. The surveyman need only set a minimal number of grade stakes to control the work. This method is illustrated in Figs. 18.2.1 and 18.2.2.

The observer can readily estimate and signal the additional depth of cut or fill required. The two strings are, of course, parallel. They need

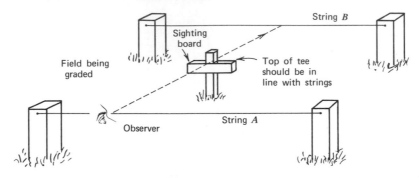

Fig. 18.2.1 String lines for grading a field.

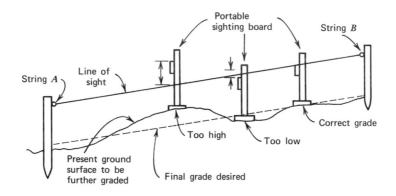

Fig. 18.2.2 Cross-section view of string-lines technique.

not be horizontal, and one may be higher than the other. This depends on the slope or gradient to be built into the construction (e.g., for natural drainage). The strings will form a plane at some convenient height above the plane of the finished ground.

18.3   Unusual Tripod Setups

Attempting to perform leveling with a 5-ft tripod in a cornfield where the corn is 8 ft high can be solved by extending the legs of the tripod with pipe or electrical conduit and sighting from atop a stepladder. (See Fig. 18.3.1.) In this situation, also, because of the row alignment of the crop, any cross sectioning for topography can be assisted if one's grid is oriented with the rows.

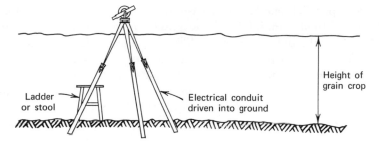

Fig. 18.3.1 Extending the tripod.

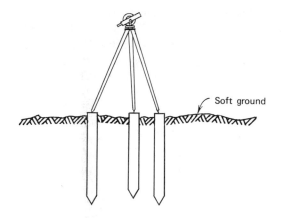

Fig. 18.3.2 Setting a transit on soft ground.

In a marshy area, where the tripod will not remain firm, a set of long stakes can be driven deep through the soft ground to establish a solid base for setting up the instrument. This is shown in Fig. 18.3.2. The same method might find an application for instrument setups in shallow water, for example, when such a setup just a bit offshore will permit good sighting up and down a shoreline without the hindrance of foliage that comes right up to the water's edge.

18.4   Measuring Inclination of Drill Holes

A transit or theodolite equipped with an optical plummet can be used to check the angle of inclination of drilled holes. For instance, as deep line drilling to form a sheer rock face when blasted needs drill holes that are very nearly parallel, one lowers a flashlight down each hole, sets up over the hole, and then manipulates the leveling screws of the transit until the light is seen in the optical plummet. Then, by means of the vertical circle and the level vial on the telescope, the angle of inclination of the hole can quickly be measured. This is an unusual but helpful maneuver worked up by a field surveyman using a new device with a bit of imagination. The procedure is sketched in Fig. 18.4.1.

18.5   The Universal Laser—Some Other Uses

The helium-neon (He-Ne) laser beam, visible even in direct sunlight, is increasingly used in construction. The illuminated spot on a reference target is available at the point where work is being done, and there is no need to peer through a transit to observe and signal to another person. Accurate laying of pipeline, setting of turbine axes, and boring of tunnels

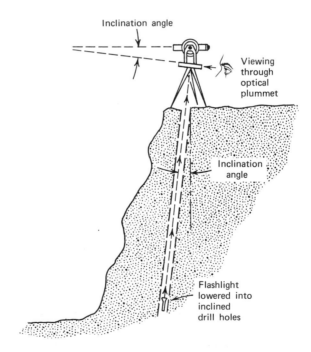

Fig. 18.4.1 Measuring drill-hole inclination by using vertical of transit.

can be done with the aid of a laser beam directed along the center line. With a photoelectric detector mounted on the tunnel-boring equipment, deviations from the alignment will actuate horizontal or vertical steering signals to realign the drilling direction. These several usages are indicated throughout, wherever the appropriate topic suggests their mention.

Other uses may suggest themselves as the laser becomes a more familiar tool. A bright laser beam has been used to make a visible target on the side of a mountain or cliff so that two transits may simultaneously observe angles to the point (horizontal and vertical) and thus discover the location and elevation of the point. Figure 18.5.1 illustrates the technique. Many points can be rapidly sighted from the same setup without great expenditure of time and effort, and the dispatching of personnel into difficult terrain.

It follows that if the laser and the transits were high on a hill, the same procedure might be used for locating points on the valley floor, again without sending rodmen from point to point. The method is a simple form of triangulation. Fig. 18.5.2 shows a transit equipped with a laser that could perform this task and other alignment tasks.

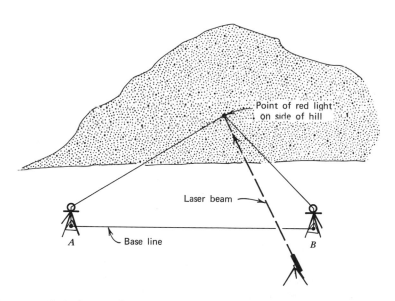

Point of red light on side of hill

Laser beam

A

Base line

B

Fig. 18.5.1 Laser beam making a spot of red light observable by two transits.

Fig. **18.5.2** Alignment laser mounted on transit.   *Courtesy Keuffel & Esser Co.*

18.6   *Some Notions on Measurement Digits*

Each measurement is a combined counting plus an estimating. The last digit is properly the estimated. Repeatability is a clue to the accuracy; for example, if the first measurement of a map measuring wheel or a scale is 31.64 in., and the next ones are 32.11, 30.92, 31.02, . . . , then one must suspect the method, or the mechanism, or the carefulness. One must be aware though, that repeatability alone is not enough; the following examples will illustrate.

1. Two tapemen can report 217.81, 217.82, 217.82, and 217.81 as four measurements with a steel tape, but various things can be wrong.
   a. They can forget this is slope distance, not horizontal.
   b. They can forget the temperature correction.
   c. They can work subsconsciously after the first reading to get the next three answers to back up the first.

d. They could even have badly goofed and the real value is 317.8± instead of 217.8±.

2. A cloth tape may have inherent stretch sufficient to provide grossly wrong although consistent answers.

3. An uncalibrated tape may have been used, and however frequently repeated, the result can be quite wrong.

18.7 Grading by the Planoscope

A new tripod-mounted optical device known as a planoscope can be set up to form virtually a visible plane to give grade lines to dozer or scraper operators. What the operator sees when his eye is in the correct plane (at the proper height) is shown in the photo of the instrument (Fig. 18.7.1). The two black bars are the clue; the schematic of Fig. 18.7.2 shows the correct, the too high, and the too low indications for the observer's eye, and these can be seen as he travels his machine back and forth along the area being graded.

With a string placed around the front, side, and back windows of his cab at just the correct height, the driver moves his head up or down until

Fig. 18.7.1 The planoscope.

CORRECT LEVEL

TOO LOW

TOO HIGH   Fig. 18.7.2 Observer's view of planoscope.

he sees a symmetrical configuration. At the same time he can observe if his sighting string is too high or too low, or whether the terrain beneath his tracks is too high or too low. The correct level has been achieved when the string can be seen to coincide with the middle of the planoscope and the two black bars match as he views them.

The instrument has more than a 90° field of view, so it can serve many pieces of equipment over a fairly extensive area. It can be set level, or can be inclined as needed. Its height can be adjusted to coincide fairly well with the height of the driver's eye as he rides the dozer or grader. The instrument, if set lower, can be used with a sighting board much as is in Section 18.2. It can also be adapted for use in trenching and laying sewer pipe to a correct gradient (as in Chapter 11), as possibly for other grading operations where the accent is on rapidity rather than on pinpoint precision.

18.8 Slope Indicator

The slope indicator is a precision device, lowered through a vertical tube into the ground, which can report any inclination from the vertical and thus detect lateral ground movements. It can be used in a deep fill to detect shear zones or unstable conditions at any depth below the surface. The detection unit is lowered by cable in a previously installed 4-in. plastic or aluminum casing having internal grooves to maintain the unit's orientation. A pendulum in the detection unit actuates a dial at the surface whenever it finds a deviation from the vertical caused by lateral earth movements. (See Fig. 18.8.1.)

An initial set of inclination readings is obtained at specific depths within the casing, and at periodic intervals of time subsequent readings are taken at the same depths. The difference in successive readings at

Fig. 18.8.1 Using the slope indicator.

identical depths represents a change in inclination of the tube, which is converted to amount and direction of linear displacement. A progressive change of readings indicates a zone of movement, as is seen from the three bore holes of Fig. 18.8.2. Chiefly the slope indicator device finds its use in such landslide and land subsidence areas, in earthfills, in earthen and rockfill dams, in areas where deep cutbanks are made, and so on. Its accuracy is surprisingly good, good enough to detect dangerous earth movements and apply corrective measures before damage occurs.

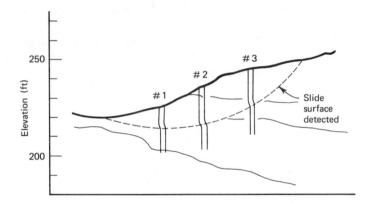

Fig. 18.8.2 Potential slide area measured by slope indicator.

19

Measurement of Areas and Volumes

19.1   Linear Measurements on Plans

For measurements on a plan or map, scales graduated to read the distances directly in feet are most convenient to use, as shown in Fig. 19.1.1. To measure the length of curved or irregular lines one can use an inexpensive map measurer, a mechanical device that reads directly in inches or centimeters. They are helpful to estimators when "taking off" quantities from plans. The line to be measured is traced with the tracing wheel as the instrument is moved along by eye (or along a straightedge or french curve). Figure 19.1.2 shows such a measurer in use.

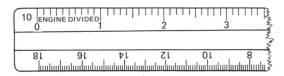

Fig. 19.1.1 Accurate scales for measurements on maps.

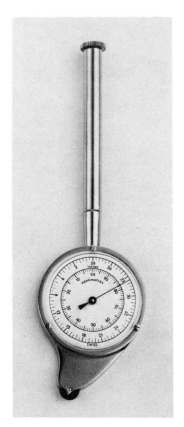

Fig. 19.1.2 A map-measuring device.

Courtesy Keuffel & Esser Co.

19.2 Area Measurements

The area of a figure is desired frequently, since it is a basis for payments or enters into a volume calculation. The purpose should be known, as a guide to the accuracy needed and thus to the tools to be employed for the measurement. Regardless, however, of the type of measurements, these methods of area calculation are general and respond to the accuracies employed.

19.2.1 By Coordinates

If coordinates of corners are known, the area is thus calculated directly (best using a desk calculator):

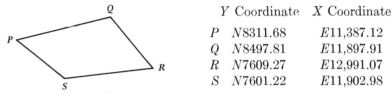

	Y Coordinate	X Coordinate
P	$N8311.68$	$E11,387.12$
Q	$N8497.81$	$E11,897.91$
R	$N7609.27$	$E12,991.07$
S	$N7601.22$	$E11,902.98$

$$A = \tfrac{1}{2}[X_P(Y_S - Y_Q) + X_Q(Y_p - Y_R) + X_R(Y_Q - Y_S) + X_S(Y_R - Y_P)]$$

(A negative result may occur, depending upon whether clockwise or counterclockwise direction is employed. Either is valid.)

X	$(Y - Y)$	Product
$+11,387.12$	-896.59	$-10,209,577.9208$
$+11,897.91$	$+702.41$	$+ 8,357,210.9631$
$+12,991.07$	$+896.59$	$+11,647,663.4513$
$+11,902.98$	-702.41	$- 8,360,772.1818$

Double area $= 1,434,524.31$

Area $= 717,262.2$ ft²

19.2.2 By DMD

If set up in traverse format, this area is calculable in virtually the same manner, using the double meridian distance (DMD) method.

Course	Latitude (ΔY)	Departure (ΔX)	DMD	Double Area
PQ	$+186.13$	$+510.79$	$+510.79$	$+95,073.34$
QR	-888.54	$+1093.16$	$+2114.74$	$-1,879,031.08$
RS	-8.05	-1088.09	$+2119.81$	$-17,064.47$
SP	$+710.46$	-515.86	$+515.86$	$+366,497.90$
			$2A =$	$-1,434,524.31$
			$A =$	$-717,262.2$

The DMD of any course is the DMD of the preceding course (if any), plus the departure of the preceding course (if any), plus the departure of the course itself. The latitude is multiplied by the DMD to get each double area, and the double area sum is divided by two. (A negative result is again not invalid.)

19.2.3 By Geometrics

If measurements are to be made from a plan or map to obtain an area, one method is to divide it into triangles, rectangles, and trapezoids and then scale the requisite dimensions for multiplication. The resultant area will be only as good as the scaling process and the correctness of the map itself.

If the boundaries are circles of known diameters and central angles, the areas of sectors or segments can be found by the appropriate formulas, and applied properly. For instance, a parcel may be as shown in Fig. 19.2.3.1.

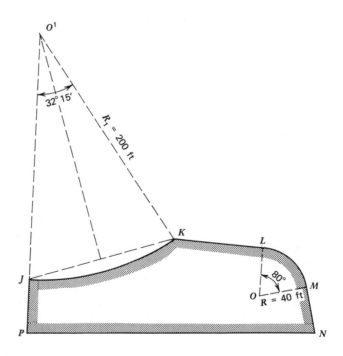

Fig. 19.2.3.1 Area by geometrics.

If the area *JKLOMNPJ* be obtained as for a straight sided figure, the segment *JK* can be subtracted and the sector *LOM* added:

For segment *JK*:

$$\text{Area} = \text{area of sector } O'JK - \text{area of triangle } O'JK$$

$$= \pi R^2 \frac{\Delta}{360} - R^2(\sin \Delta/2 \cos \Delta/2)$$

$$= \pi(200)^2 \left(\frac{32.25}{360}\right) - (200)^2 (0.22773) (0.96066)$$

$$= 11{,}257 - 8751 = 2506 \text{ ft}^2 \text{ (to be subtracted)}$$

For sector *LOM*:

$$\text{Area} = \pi R^2 \frac{\Delta}{360} = \pi(40)^2 \left(\frac{80}{360}\right)$$

$$= 1117 \text{ ft}^2 \text{ (to be added)}$$

19.2.4 By Counting Squares

A quick graphical method of measuring for area of an irregular figure is to overlay the area with a transparent grid and count the squares within the area. It helps to have the "tenth lines heavy" to speed the process, since large blocks of squares can be rapidly counted this way. (See Fig. 19.2.4.1.)

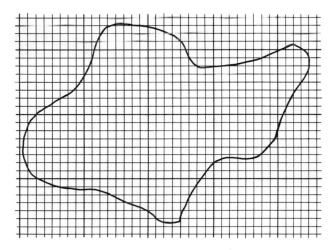

Fig. 19.2.4.1 Graphical area measurement by counting squares.

19.2.5   By Planimeter

A map or plan area can also be measured directly by a planimeter, a device that gives the area enclosed as the point moves about the boundary. A polar planimeter is shown in Fig. 19.2.5.1. By simply moving the tracer pin or lens around the periphery of the figure, the area is read directly from the measuring wheel and dial; the method is far simpler and far more accurate than counting squares.

Making repeated measurements of the area, allowing the planimeter areas to accumulate on the scale, and then dividing by the number of times will increase precision—all other things being equal—in the same manner that an angle is measured more precisely by the method of repetition.

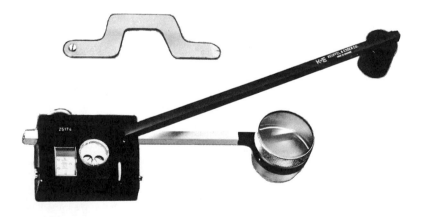

Fig. 19.2.5.1 Polar planimeter for measuring map areas.   *Courtesy Keuffel & Esser Co.*

19.3   Use of the Polar Planimeter

The polar planimeter measures area by having the tracing point follow the outline of the irregular area. The pole of the planimeter usually rests outside the area, care being taken to have the arms as nearly at right angles as possible. If they get further than 45° away from that, it is better to break up the area into two or more portions and do each subarea separately, moving the pole each time to a new location. The location of the pole is arbitrarily chosen but remains fixed throughout any one area measurement. Figure 19.3.1 shows a polar planimeter measuring a map area. With all planimeters, the movement of the tracer point around the periphery of the figure causes the measuring wheel to revolve, the amount of revolution depending on the distance moved and the angle formed by the axis of the measuring wheel and its direction of motion. If the tracer is moved clockwise, the readings will increase; if counterclockwise, they will decrease.

In the example shown, if the irregular area were part of a map at scale 1 in. = 2000 ft, the area in square feet would be 83.96 sq in. multiplied by the scale squared, or:

$$A\,(\text{ft}^2) = 83.96 \text{ in.}^2 \times \frac{2000 \text{ ft}}{1 \text{ in.}} \times \frac{2000 \text{ ft}}{1 \text{ in.}}$$

$$= 335,840,000 \text{ ft}^2$$

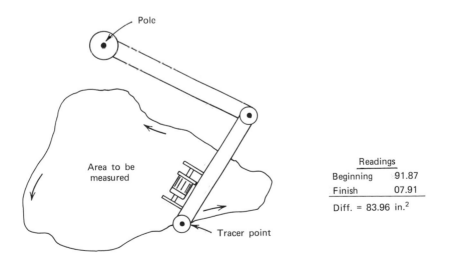

Readings	
Beginning	91.87
Finish	07.91
Diff. = 83.96 in.^{2}	

Fig. 19.3.1 Polar planimeter being used to find the area of irregular figure.

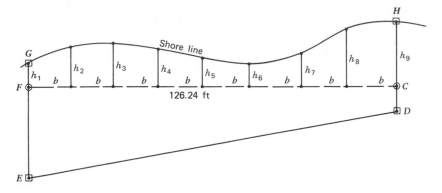

Fig. 19.4.1 Area bounded by an irregular side.

19.4   The Trapezoidal Rule

An area with an irregular curved boundary (as along a shore) can be calculated by measuring multiple parallel offsets to the curved boundary from a reference line. The number of offsets is determined by the degree of irregularity of the curved boundary and the degree of accuracy required. The reference line must be perpendicular to the offsets. The area *FGHCF* of Fig. 19.4.1 is calculated by the trapezoidal rule as an example; the area *CDEFC* can be found by other means.

$$A = \tfrac{1}{2}b \left[h_1 + h_n + 2(h_2 + h_3 = \ldots + h_{n-2} + h_{n-1}) \right]$$

If $b = 15.78$ ft, and the measured offset values are as shown,

h	ft	h	ft	h	ft
1	8.91	4	13.01	7	13.07
2	13.22	5	12.22	8	18.39
3	13.19	6	11.92	9	21.31

the calculated area is

$$A = \tfrac{1}{2}(15.78) \, [8.91 + 21.31 + 2(95.02)] = 1738 \text{ ft}^2$$

This is exactly the sum of the areas of the small trapezoids shown, and assumes that the shore line is straight between offsets.

19.5   Simpson's One-Third Rule

This rule is based upon a curved line, rather than on a succession of straight lines, and tends to give closer results. It requires an odd number of offsets:

$$A = \frac{b}{3}[h_1 + h_n + 2(\Sigma\, h_{\text{other odd}}) + 4(\Sigma\, h_{\text{even}})]$$

In this case, again referring to Fig. 19.4.1,

$$A = \frac{15.78}{3}[8.91 + 21.31 + 2(38.48) + 4(56.54)] = 1753 \text{ ft}^2$$

The area obtained by Simpson's One-Third Rule is a much more correct value than that found by the trapezoid rule.

19.6   Three-Level Cross Sections

Dikes, berms, or route embankments that are level across the top can be simply calculated by dividing them into suitable triangles. The same principle applies to cut sections (see Section 2.20.). To calculate the area of a cut or a fill cross section of the three-level type—having dimensions for only the center point and the two side slope-stake points—the formula referring to Figs. 19.6.1 and 19.6.2, is:

$$A = \frac{c}{2}(d_L + d_R) + \frac{\frac{1}{2}b}{2}(h_L + h_R)$$

One can see the triangles involved by drawing the long diagonal (shown dashed).

As an example of the area computation of a section, suppose that the

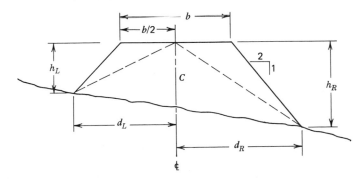

Fig. 19.6.1 Three-level cross section in a fill or embankment. The side slope is shown here as 2:1.

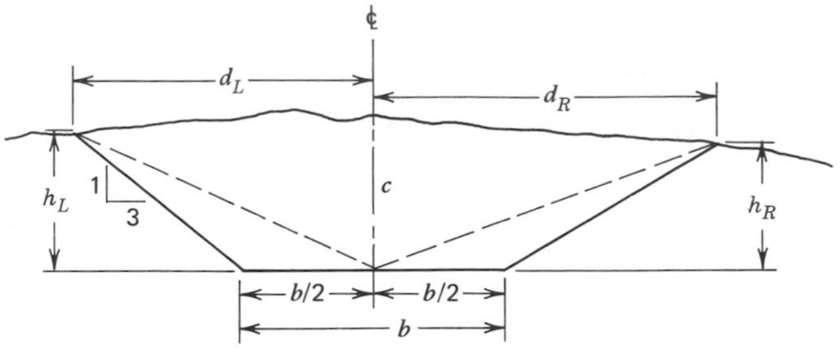

Fig. 19.6.2 Three-level cross section in cut. The side slope is shown here as 3:1.

cross-section notes for Sta. 312 resulting from the slope-staking process of Section 2.20 are as follows:

Sta.	L	C	R	$b = 22$ ft
$312 + 00$	-6.8	-8.2	-11.6	slope is $2:1$
	$\overline{24.6}$	$\overline{0.0}$	$\overline{34.2}$	

The negative signs indicate a fill (embankment); if this were a cut, the signs would be positive. These notes correspond to Fig. 19.6.1. Identification of the values is given for clarity:

$$h_L = 6.8 \qquad\qquad C = 8.2$$
$$h_R = 11.6 \qquad\qquad b = 22.0$$
$$d_L = 24.6 \qquad\qquad \text{Side slope} = 2:1$$
$$d_R = 34.2$$

The area can be calculated from the formula.

$$A = \frac{-8.2}{2}(24.6 + 34.2) + \frac{\frac{1}{2}22}{2}(-6.8 - 11.6)$$

$$= -4.1\,(58.8) + 5.5\,(-18.4)$$

$$= -241.1 - 101.2 = -342.2 \text{ ft}^2$$

Normally, with a large number of these area computations (along with volume calculations as in Section 20.2), the work is tabulated in columns for convenience and speed.

19.7   Multilevel Cross Sections

Sometimes a five-level cross section is used, giving more values to be computed, as a refinement of the three-level section. Figure 19.7.1 shows

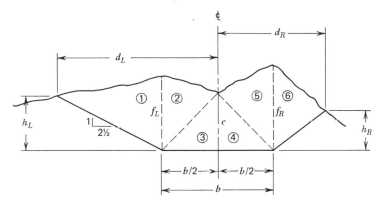

Fig. 19.7.1 A five-level cross section in cut. The side slope is shown here as 2-1/2:1.

such a section; it is apparent that the two extra depth readings occur at the edge of the base width. Typically, the slope-stake notes for a sample section might be:

Sta.	L		C		R	
507 + 00	+7.2	+13.1	+9.6	+14.4	+3.6	b = 20 ft
	28.0	10.0	0.0	10.0	18.9	Slope $2\frac{1}{2}$:1

The identification of values is as given in the preceding article except that now there are two others:

$$f_L = +13.1 \qquad\qquad f_R = +14.4$$

A study of Fig. 19.7.1 shows that the area of the five-level section is worked up from triangles, as follows:

$$\text{Triangles 1 and 2: } A = \tfrac{1}{2}d_L f_L$$

$$\text{Triangles 3 and 4: } A = \tfrac{1}{2}bC$$

$$\text{Triangles 5 and 6: } A = \tfrac{1}{2}d_R f_R$$

$$\text{Total area} = \tfrac{1}{2}(bC + d_L f_L + d_R f_R)$$

$$A = \tfrac{1}{2}[(20 \times 9.6) + (28.0 \times 13.1) + 18.9 \times 14.4)]$$

$$= \tfrac{1}{2}(192.0 + 366.8 + 272.2) = \tfrac{1}{2}(831.0) = 415.5 \text{ ft}^2$$

Where the original ground surface is very irregular and the cross section does not lend itself to being worked as a three-level or a five-level section, an "irregular" multilevel section occurs. Such a section is given

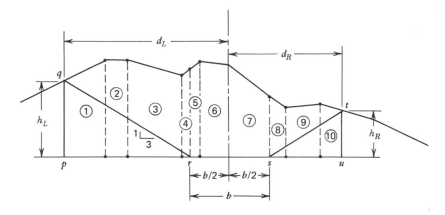

Fig. 19.7.2 Multilevel cross section in cut.

here only as a sample of what can occur. It is shown in Fig. 19.7.2, with the dashed lines indicating the trapezoids that can be calculated and summed, and the two side triangles that are then subtracted to give the net area. There are 10 trapezoids in this example; the two triangles are *p q r* and *s t u*.

A similar multilevel cross section is shown in Fig. 19.7.3, with the elevation of each point written above the point's distance to right or left of the center line. The computation of such an area is best made by the

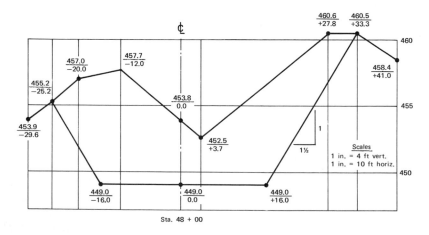

Fig. 19.7.3 Multilevel cross section. The base is 32 ft and the side slope is 1-1/2:1.

coordinate method (see Section 19.2.1). The area of Fig. 19.7.3 is calculated here as an example.

$$Y_n(X_{n-1} - X_{n+1}) \qquad = \text{double area}$$

$$449.0\,(-16.0 - 16.0) = 449.0\,(-32.0) = -14{,}368.0$$

$$449.0\,(-25.2 - 0.0) = 449.0\,(-25.2) = -11{,}314.8$$

$$455.2\,(-20.0 + 16.0) = 455.2\,(-4.0) \quad - \quad -1{,}820.8$$

$$457.0\,(-12.0 + 25.2) = 457.0\,(+13.2) = +6{,}032.4$$

$$457.7\,(0.0 + 20.0) \quad = 457.0\,(+20.0) = +9{,}140.0$$

$$453.8\,(+3.7 + 12.0) = 453.8\,(+15.7) = +7{,}124.7$$

$$452.5\,(+27.8 - 0.0) = 452.5\,(+27.8) = +12{,}579.5$$

$$460.6\,(+33.3 - 3.7) = 460.6\,(+29.6) = +13{,}633.8$$

$$460.5\,(+16.0 - 27.8) = 460.5\,(-11.8) = -5{,}433.9$$

$$449.0\,(0.0 - 33.3) \quad = 449.0\,(-33.3) = -14{,}951.7$$

$$2A = \text{Sum} = +48{,}510.4 - 47{,}889.2 = +621.2$$

$$A = +310.6 \text{ ft}^2$$

Were the area of this same multilevel cross section to be computed by trapezoids and triangles, these shapes and dimensions would be as shown in Fig. 19.7.4 and the area computation would be as shown in Table 19.7.1.

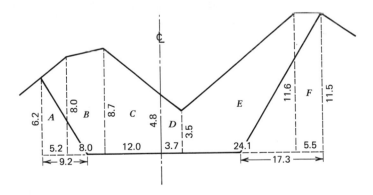

Fig. 19.7.4 Multilevel cross section showing trapezoid and triangle dimensions for area computation.

Table 19.7.1   Area of Multilevel Section

Trapezoids	Heights	Mean	Width	Area	Net
	6.2				
A		7.1	5.2	36.9	
	8.0				
B		8.4	8.0	67.2	
	8.7				
C		6.8	12.0	81.6	
	4.8				
D		4.2	3.7	15.5	
	3.5				
E		7.6	24.1	183.2	
	11.6				
F		11.6	5.5	63.8	
	11.5				+448.2

Triangles	Area				
Left	½ (6.2 × 9.2) =			28.5	
Right	½ (11.5 × 17.3) =			99.5	−128.0
Total area of cross section	=				+320.2 ft²

Where the finished cut or fill surface, the new "grade," is not level and/or not straight, a more complex situation arises. Figure 19.7.5 shows the cross section of a four-lane divided highway at a typical station, showing the shape of the finished earthwork prior to the paving or finished grade. This cross-section configuration thus represents the subgrade, and the term "templet" or "subgrade templet" designates the shape of roadway, shoulders, median strip, ditches, and side slopes. The seem-

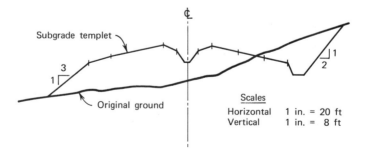

Fig. 19.7.5 Multilevel side-hill cross section with subgrade templet for divided four-lane highway. Side slope in cut is 2:1, and in fill 3:1.

ingly exaggerated shape results from using different scales for horizontal and vertical for the sake of clarity.

To calculate the cut and fill areas of such an irregular cross section, triangles and trapezoids could possibly be used, but it is more likely that the area would be calculated by the coordinate method. Although not shown on the schematic drawing of Fig. 19.7.5, the elevation and distance from center line would be available for each point for use in the calculation.

An extremely simple procedure is to use a strip of paper, as in Fig. 19.7.6, to measure cumulatively the depths at regular intervals across the cross section (or to use a map measurer, as in Section 19.1). Then the total accumulation is divided by the number of verticals measured to get the average depth; the average depth in feet multiplied by the total width of the section in feet will give the area of the cross section. In any side-hill section, for example, Fig. 19.7.5, where there is both cut and fill, the two area values must be distinguished, with all cut $(+)$ and all fill $(-)$ tallied separately.

If the cross section is drawn to scale, the planimeter can be used to get the area directly in square inches. Then, because cross sections of highways are usually drawn with vertical scale exaggerated, the conver-

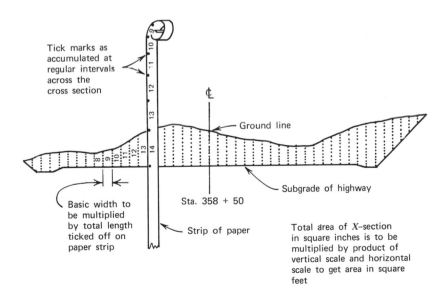

Fig. 19.7.6 Highway cross-section area found by use of paper strip on which depths of cut are ticked off in a cumulative fashion across the diagram.

sion from square inches (on paper) to square feet (actual area) is accomplished by use of two different scale multipliers. For instance, if the horizontal scale is 1 in. = 10 ft and the vertical scale is 1 in. = 4 ft, the planimeter value (in.^{2}) is multiplied by 40 (= 10 × 4) to give the area (ft^2).

19.8   Invisible Cross Sections

As distinct from the cross sections just described, methods have been developed to "draw" the ground profile in the memory of a computer or on punched cards or tape, working directly from photogrammetric equipment where the operator sees a stereoscopic model of the terrain. No actual cross section is drawn. Instead, the stereo operator moves along the route and at every station (or half-station) sets his floating point marker on the ground in the model. Then the values of x-coordinate, y-coordinate, and z-coordinate (elevation) are read by electric sensors and placed in storage. The operator moves his mark similarly to right and to left of the center line to get x, y, and z coordinates at each significant point, thus doing a cross section at each station.

Subsequently a computer program calculates the area of the cross section by interacting this "ground" information with the "grade" information (the templet in mathematical form). There is then no need to draw the profile of the route or the cross-section views on paper, a significant saving in time and effort. In fact, the computer program goes a step further and calculates volumes of cut and fill also. There are a number of computer variations, but omitting to draw cross sections of highways and storing them invisibly for the computer is becoming commonplace.

20

Volume Computations

20.1   Earthwork Volumes

Calculation of volumes, sometimes called solidities, is accomplished in a variety of ways. Since volume is based on area, many of the preceding area measurement methods find a usage in computing volumes of earthwork in dredging, trenching, digging canals or drainage ditches, constructing dikes, or doing cut or fill in highway work. A great amount of earth moving is done for highway building, and the cross-section areas previously discussed form a basis for its calculation. Volumes or solidities are calculated from cross-section areas of the cuts, side-hill sections, and fills, with continuing cumulative solidities. These calculations are commonly done by computer, basically by this method, with printouts at each station or more frequently. Illustrative examples are given of the methods of volume calculation, which are basically alike whether done by hand, by desk calculator, or by computer.

20.2   Average End-Area Method

For highways, dikes, dams, channels, or other earthwork, the simple method of multiplying the length between two cross sections by the

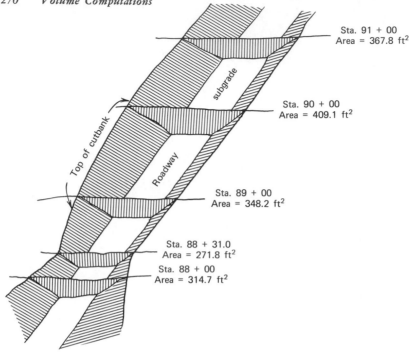

Sta. 91 + 00
Area = 367.8 ft²

subgrade

Sta. 90 + 00
Area = 409.1 ft²

Top of cutbank

Roadway

Sta. 89 + 00
Area = 348.2 ft²

Sta. 88 + 31.0
Area = 271.8 ft²

Sta. 88 + 00
Area = 314.7 ft²

Fig. 20.2.1 Portion of a highway in cut, showing cross sections.

average of the two cross section areas will usually give a satisfactory value for the volume. This number serves as an advance estimate for calculating the extent of the project, and also as a basis for payment for the work done.

Figure 20.2.1 shows a portion of a highway in cut, with the cross-sectional areas as found by measurement. The volume of earthwork between the section shown can be found by the formula

$$V = \frac{L}{27}\left(\frac{A_1 + A_2}{2}\right)$$

where

$$V = \text{volume (cu yd)}$$

$$A_1, A_2 = \text{cross-sectional areas (sq ft)}$$

$$L = \text{length between the sections (ft)}$$

$$27 \text{ is used to convert cu ft to cu yd}$$

The volume between Sta. 88 + 00 and 88 + 31.0 is

$$V = \frac{31}{27} \left(\frac{314.7 + 271.8}{2} \right) = +336.7 \text{ yd}^3$$

and between Sta. 90 + 00 and 91 + 00,

$$V = \frac{100}{27} \left(\frac{409.1 + 367.8}{2} \right) = +1438.0 \text{ yd}^3$$

Tabulation techniques can reduce the work to a simpler format; the calculation in Table 20.2.1 of the entire highway segment illustrates such a simplified procedure.

Table 20.2.1   Volume Computation by Average End-Area Method

Sta.	Length (ft)	Area (ft^2)	Mean Area (ft^2)	Volume (yd^3)	Cumulative Volume (yd^3)
88 + 00		+314.7			−1762.8[a]
	31.0		+293.2	+336.7	
88 + 31.0		+271.8			−1426.1
	69.0		+310.0	+792.1	
89 + 00		+348.2			−634.0
	100.0		+378.6	+1402.1	
90 + 00		+409.1			+768.1
	100.0		+388.4	+1438.0	
91 + 00		+367.8			+2206.1

[a] This value is the cubic yardage accumulated from previous stations, which has included a greater amount of fill than of cut, as the negative sign indicates.

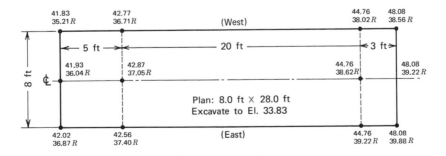

Fig. 20.2.2 Plan view of fuel tank excavation in earth and in rock.

The excavation for installing an underground fuel tank depicted in Fig. 20.2.2 was measured by elevation readings on the original surface, then on the rock surface as it was exposed. The cylindrical tank, 26 ft long and 6 ft in diameter, required a flat base for cradles, or a rectangular outline 8 x 28 ft (to include 1 ft extra to the payment line on each face), and needed 1 ft of cover. The original ground elevations and the

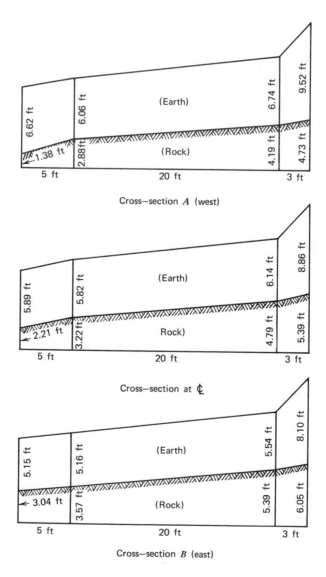

Cross—section *A* (west)

Cross—section at ℄

Cross—section *B* (east)

Fig 20.2.3

elevation of rock are shown on the diagram; excavation was carried to elevation 33.83 ft, or 8 ft below the lowest ground point.

The areas for earth and for rock of the cross sections shown in Fig. 20.2.3 can be computed as the sums of the areas of trapezoids, or can be measured by planimeter if drawn to scale.

By use of the average end-area method, the volumes of excavation of earth and rock for the fuel tank of Fig. 20.2.2 are also calculated and appear as Table 20.2.2. The cross sections are shown in Fig. 20.2.3, with dimensions taken from the results of leveling during the work. Earth and rock excavations are calculated separately, since the unit prices are quite different.

Table 20.2.2a   Fuel Tank Excavation Volumes by Average End-Area Method: Earth Quantities

Trapezoid	Height (ft)	Average (ft)	Width (ft)	Area (ft²)	Section Area (ft²)	Mean Area (ft²)	Length (ft)	Volume (yd³)
	6.62							
A-1		6.34	5.0	31.7				
	6.06							
A-2		6.40	20.0	128.0				
	6.74							
A-3		8.13	3.0	24.4				
	9.52				184.1			
						177.8	4.0	26.3
	5.89							
C-1		5.86	5.0	29.3				
	5.82							
C-2		5.98	20.0	119.6				
	6.14							
C-3		7.50	3.0	22.5				
	8.86				171.4			
						162.4	4.0	24.0
	5.15							
B-1		5.16	5.0	25.8				
	5.16							
B-2		5.35	20.0	107.0				
	5.54							
B-3		6.82	3.0	20.5				
	8.10				153.3			

Total volume of earth excavation = 50.3 yd³

Table 20.2.2b   Fuel Tank Excavation Volumes by Average
End-Area Method: Rock Quantities

Trapezoid	Height (ft)	Average (ft)	Width (ft)	Area (ft²)	Section Area (ft²)	Mean Area (ft²)	Length (ft)	Volume (yd³)
	1.38							
A-1		2.13	5.0	10.6				
	2.88							
A·2		3.54	20.0	70.8				
	4.19							
A-3		4.46	3.0	13.4				
	4.73				94.8			
						101.8	4.0	15.1
	2.21							
C-1		2.72	5.0	13.6				
	3.22							
C-2		4.00	20.0	80.0				
	4.79							
C-3		5.09	3.0	15.3				
	5.39				108.9			
						116.1	4.0	17.2
	3.04							
B-1		3.30	5.0	16.5				
	3.57							
B-2		4.48	20.0	89.6				
	5.39							
B-3		5.72	3.0	17.2				
	6.05				123.3			

Total volume of rock excavation = 32.3 yd³

20.3   Prismoidal Method

A refinement in volume calculating is the prismoidal formula, which gives
a more accurate result than the average end-area method. This is especially true whenever the two end areas are not nearly of the same magnitude. The prismoidal formula is:

$$V = \frac{L}{6} (A_1 + 4 A_M + A_2) \ldots \text{answer in cu ft}$$

or

$$V = \frac{L}{6 \times 27} (A_1 + 4 A_M + A_2) \ldots \text{answer in cu yd}$$

where

$$V = \text{volume (cu ft or cu yd)}$$

$$A_1, A_2 = \text{cross-sectional areas of the two end sections (sq ft)}$$

$$A_m = \text{cross-sectional area of the middle section (sq ft)}$$

An illustration is provided in calculating the amount of concrete required to build the pedestal for a statue (Fig. 20.3.1). By the prismoidal formula,

$$V = \frac{5}{6} [4 + 4(16) + 36] = 86.67 \text{ ft}^3 = 3.21 \text{ yd}^3$$

An erroneous answer would be given by the average end-area formula in this case, since the two end areas differ so greatly,

$$V = 5 \left(\frac{4 + 36}{2} \right) = 100 \text{ ft}^3 = 3.71 \text{ yd}^3$$

The prismoidal formula gives a more correct answer if used in highway volume calculations, and is used in isolated instances. For example, using the area values of the last 200 ft shown in Table 20.2.1, the prismoidal formula gives as the volume:

$$V = \frac{200}{6 \times 27} [+348.2 + 4(+409.1) + 367.8] = 2904.2 \text{ yd}^3$$

as opposed to 2840.1 yd³ (= 1402.1 + 1438.0 from the table)

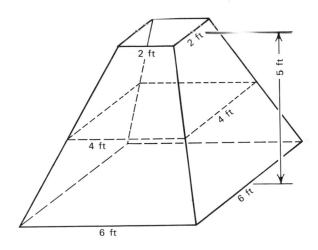

Fig. 20.3.1 Concrete pedestal.

Prismoidal correction factors do exist for three-level sections in earthwork and highway manuals, although the refinement introduced does not usually bring about any great cost saving, and the precision may not be justified because it could easily be superior to that of the field measurements made to secure earthwork data.

20.4   Borrow-Pit Method

By the grid-type or range lines of Section 2.23, the manner of determining ground elevations was explained. If these elevations are recorded and the grid intersections are later reestablished, any excavation (or filling) can be measured, for payment purposes. A new gridiron cross sectioning on the original points will furnish new elevations, and the depth of cut at each point can be ascertained. The computation of the volume of material removed can be determined as follows.

Approximate means of aligning the rodman may sometimes suffice, such as having him sight himself onto a range line and pace off his distances. In difficult terrain, however, where a small lateral displacement will give an appreciable distortion of elevation, there may be need even to stake out the grid squares again. Certainly if rock excavation is to be measured, always an expensive item, extreme care must be employed, as a safeguard against future controversy, as to quantities removed. The "borrow-pit method" of computing the volumes of earth and/or rock removed requires the comparison of "before" and "after" elevations at each grid intersection. Thus the depths of cut at each corner are tallied.

The depths of cut at the four corners of each square are averaged, then multiplied by the area of the square to obtain the volume (of the prism). Because any particular corner may figure in as many as four squares (or prisms), a quicker volume computation is feasible. The corners involved in one prism only are summed; those involved in two prisms only, or in three, or in four are separately summed. Then the volume of the entire excavation is found thus:

$$\text{Volume} = \frac{\text{Area of one square}}{4} \times \left\{ \begin{array}{l} \text{Sum of height figuring in one} \\ \text{prism only} + \text{twice the sum of} \\ \text{height figuring in two prisms} \\ + \text{three times the sum of heights} \\ \text{figuring in three prisms} + \text{four} \\ \text{times the sum of heights} \\ \text{figuring in four prisms.} \end{array} \right\}$$

As an example, the set of notes for a sample borrow pit are shown in Fig. 20.4.1, and the calculations are as follows:

$$V = \frac{10' \times 10'}{27 \times 4} [(60.5) + 2(81.3) + 3(19.6) + 4(51.6)]$$

$$= \frac{10 \times 10}{27 \times 4} (489.5) = 453.2 \text{ yd}^3$$

Where it is desirable to ascertain quantities of both excavated earth and excavated rock, the initial elevation readings are made on the ground surface, then at each grid intersection on the surface of the rock as it is exposed, and finally at the finish of the excavation. The quantity of rock is calculated, and the total quantity of both earth and rock is calculated. Subtracting one from the other gives the volume of earth excavation. This is a normal way to measure excavation quantities for payment.

POINT	ORIG. ELEV.	FINAL ELEV.	DEPTH	Nº OF SQUARES
A 1	93.5	81.5	12.0	1
2	93.1	81.5	11.6	2
3	92.8	81.5	11.3	1
B 1	92.9	81.5	11.4	2
2	92.7	81.5	11.2	4
3	92.3	81.5	10.8	2
C 1	92.5	81.5	11.0	2
2	92.2	81.5	10.7	4
3	91.8	81.5	10.3	3
4	91.4	81.5	9.9	1
D 1	92.0	81.5	9.5	2
2	91.7	81.5	10.2	4
3	91.4	81.5	9.9	4
4	89.9	81.5	8.4	2
E 1	91.5	81.5	10.0	2
2	91.1	81.5	9.6	4
3	90.8	81.5	9.3	3
4	90.5	81.5	9.0	1
F 1	90.9	81.5	9.4	1
2	90.7	81.5	9.2	2
3	90.4	81.5	8.9	1

BORROW-PIT LEVELING: REID, INC. BUILDING.

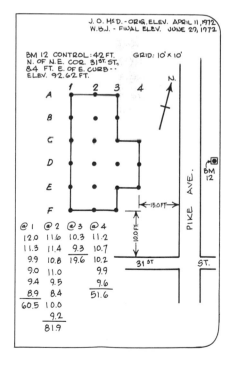

Fig. 20.4.1 Notes for borrow-pit leveling.

20.5   Contour-Slice Method

If areas can be measured by a planimeter on a contoured plan (or topographic map, as in Appendix D), it is possible to use a variation of the average end-area method to find the volume of earth work. This contour-slice method supposes that the area confined within a particular contour is a closed area on the plan. Successive contour areas, then, will form a sort of "slice" whose thickness is the contour interval, vertical distance between successive contoured areas. The average of two successive areas multiplied by this contour interval will give the volume of material, whether earth, rock, coal, iron ore, or water.

To illustrate, a contoured map is shown in Fig. 20.5.1 with the dam to be built at the spot shown to impound water falling on the watershed area. The watershed is determined by the ridgeline around the basin and is marked by the dashed line.

The watershed area is measured on the map by use of a planimeter, giving 18.205 in.^{2}; since from the scale of 1 in. = 500 ft it is seen that a 1 in.^{2} on the map equals 250,000 ft^2 on the ground, the watershed area converts to 4,551,300 ft^2 on the ground, about 104.5 acres. Multiplying this area by the depth of anticipated rainfall (30 in.) gives 11,378,200 ft^3 or 85.109 million gallons annually (because 1 ft^3 = 7.48 gallons); it can also be stated as a volume of 261.2 acre-ft (= 104.5 acres $\times$ 2.5 ft).

The spillway elevation at the dam is to be set at elevation 94, which determines the maximum extent of the reservoir when it is full: the shoreline will be the 94-ft contour line. On the plan the area within each contour line is measured by planimeter, each area being the area of the lake surface as the water reaches that elevation while the reservoir is filling. Converting the map areas (in.^{2}) to ground areas (ft^2) requires the use of the map scale; 1 in.^{2} on the map equals 500 $\times$ 500 or 250,000 ft^2 on the ground. Successive slices of water 10 ft thick, whose top and bottom areas are averaged to give a mean area, are then computed for volume; the mean area times the depth gives the volume in cubic feet (then converted to gallons). The cumulative volume in gallons is also found, to show the total volume in the reservoir when the surface is at the elevation indicated. Table 20.5.1 shows the calculations.

It should be apparent that by the contour-slicing method volumes of earthwork (cut or fill) can be similarly measured in construction work. The volumes of coal piles are regularly found in this way for monthly inventory by utilities and other heavy users of coal, with contours being drawn from photographs taken by overflying aircraft. The slice procedure is a simple extension of the average end-area method of measuring volume.

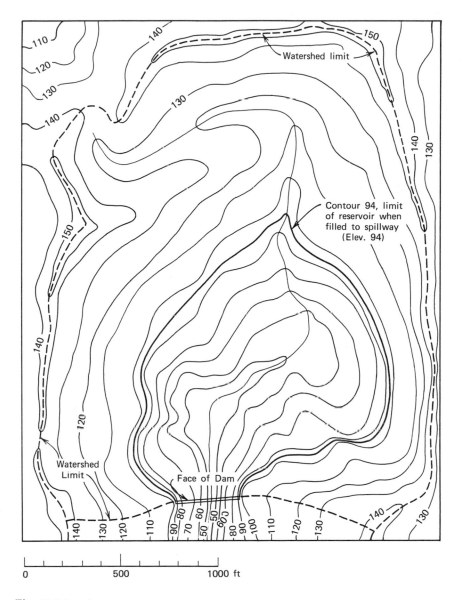

Fig. 20.5.1

279

Table 20.5.1   Reservoir Volume by Contour-SliceMethod

Contour	Area (in.²)	Area (ft²) ÷1000	Mean (ft²) ÷1000	h (ft)	Volume (ft³) ÷1000	Volume (million gallons)	Cumulative volume (million gallons)
			Reservoir (within enclosed contours)				
50[a]	0.418	104.4					0.000
			224.1	10	2241.0	16.763	
60	1.375	343.8					16.763
			470.7	10	4707.0	35.208	
70	2.390	597.6					51.971
			747.0	10	7470.0	55.876	
80	3.586	896.4					107.847
			1034.1	10	10341.0	77.351	
90	4.687	1171.8					185.198
			1237.5	4	4950.0	37.026	
94	5.213	1303.2					222.224
			Watershed (within dashed line)				
	18.205	4551.3		2.5[b]	11378.2	85.109	

[a] Water below El. 50 not considered.

[b] Annual rainfall of 30 in. estimated.

Time required to fill reservoir $= \dfrac{222.224 \text{ m.g.}}{85.109 \text{ m.g./yr}} = 2.61 \text{ yr.}$

Appendices

Special Problems, Special Topics

A

Units of Measurement

Because much attention is currently being given to metrication in the United States, and because there has been a recent change (1959) in the definition of the inch and agreement thereon among all the nations in the world still using it, a few words are added here to give the current status of our length units. Also, we have a new term, the "U.S. Survey Foot," which needs explaining. What follows will suffice to straighten out these notions.

Length

1 statute mi = 5280 ft

1 nautical mi = 6076.11549 international ft (since 1959)

$\quad\quad\quad\quad$ = 1852.0 m (since 1959)

$\quad\quad\quad\quad$ = 2025.3718 yd (since 1959)

1 yd = 3 ft = 36 in. = 0.9144 m (since 1959)

1 ft = 12 in. = 2.54 cm (exactly) (since 1959) = 0.3048 m (exactly)

3.2808399 ft = 1 m (since 1959)

When in 1959 the Foot System was redefined by agreement among officials of the nations where it is used, the inch was defined as 2.54 cm exactly, the foot as 30.48 cm exactly, and the yard as 91.44 cm exactly. This generally reduced the lengths of U. S. Foot System length units by about two parts in a million.

However, because it would have caused a great revision of existing position and elevation records of United States surveys, it was decided to retain the old definition of foot for surveys (based on the relationship

39.37 in. = 1 m exactly), and call this the U. S. Survey Foot. Thus, this exception does still exist and applies to horizontal and vertical surveying information. All information and records furnished by the U. S. Coast and Geodetic Survey and its successor, the National Geodetic Survey, is based on the U. S. Survey Foot. This is not a serious matter, since all nation-wide control surveys done by the U. S. Coast and Geodetic Survey in the past have been worked in the metric system, with conversion to the foot solely for the publication of data. Tie-in surveys, in feet, worked locally from the higher order control marks, add only small discrepancies.

The U.S. Survey Foot continues to be based on the previous value of the 1954 agreement that 1 yd = 39.37 in. or 1 nautical mi = 6076.10333 U.S. ft = 1852.0 m, which would give these relationships:

$$1 \text{ yd} = 0.91440183 \text{ m}$$

$$1 \text{ in.} = 2.54000508 \text{ cm}$$

$$1 \text{ ft } = 0.304800610 \text{ m (approx.)} = \frac{1200}{3937} \text{m (exactly)}$$

$$3.2808333 \text{ ft} = 1 \text{ m}$$

These values relate only to the U.S. Survey Foot and will continue to exist until it becomes desirable and expedient to readjust the basic geodetic survey networks in the United States, after which the ratio, as implied by the international yard, shall apply.

In summary, the differences are apparent from:

$$\text{The "new" foot (since 1959)} = 0.3048 \text{ m (exactly)}$$

$$\text{The U.S. Survey foot} = \frac{1200}{3937} \text{ (exactly)} = 0.304800610 \text{ m}$$

B

The Laser

The word *laser* is an acronym standing for "Light Amplification by Stimulated Emission of Radiation." The laser is rapidly becoming a new tool in construction alignment, as well as in distance measurement. A simple explanation of the laser principle, together with some suggestions of its versatility, is given here, and it is discussed throughout the text wherever applicable. New applications can be anticipated beyond those given in the text; optimistic estimates are that the laser will soon be virtually the only tool needed on the construction site for layout and measurement.

How the Laser Works

The laser is a light beam but, unlike ordinary light that contains the whole spectrum (red, orange, yellow, green, blue, and violet), the laser contains only one color or one wavelength of light. It thus has a helpful characteristic: it is coherent. This means that it does not tend to scatter, that the rays in the beam tend to stay almost parallel and not diverge like ordinary light.

Although laser beams were originally generated from solid crystals, today most are of the gas type. Helium and neon, contained in a cylindrical tube, are excited ("stimulated") by the passage of an electric current to form a neon glow.

By means of mirrors at the ends of the cylinder, the light is reflected back and forth, gaining in intensity by the stimulation of other atoms in the gas until the beam is so intense as to break through one mirror (which is only a half-silvered mirror, or "partial" mirror) and projects as a continuous light beam. It is this red light beam of 6328 Å (angstroms) that is visible and that remains straight without scattering. Figure B.1 shows a simplified laser-generating tube.

Lasers in common use are not expensive, not strong enough to be harmful, and should not be confused with those much more powerful lasers that can cut through steel. Alignment lasers run at about 2 or 3 mW output, almost completely harmless, although one should not look directly into the beam any more than one should look directly at the sun; either can do some damage to the retina of the eye.

The usual laser in construction does, of course, diverge somewhat with distance—and thus becomes less intense. At 1000 ft the ray can become up to $1\frac{1}{2}$ in. in diameter and under daylight conditions is difficult to see on a target. However, such a situation would not normally be encountered in practice; if warranted, a more intense laser can be used.

It must also be pointed out that laser light will follow the laws of refraction of ordinary light (see Sections 2.1 and 17.6). Long laser sights to set grade should be avoided whenever possible, or else the correction formula must be applied.

Reference is made in the text to other characteristics of laser light, that it can be reflected by a mirror or a retroprism, or "formed" into a vertical or horizontal plane by a special lens, and so on. The fan lens attachment is simply clamped on in front to convert the line of light to a plane of light, either horizontal or vertical. A fan laser can be turned horizontal and slightly inclined for giving level to graders working on a field that slopes slightly. Putting it at the height of the dozer operator's eye enables him to glimpse when he is low or high as he rides along. (In bright sun it may not readily be seen shining onto the rod or onto the excavating equipment if formed, but the operator's eye can pick it up.)

Laser light is also not disturbed by wind, rain, or ocean spray, although it will not penetrate fog. A light mist can generally be penetrated by the laser, however, so that work can continue under conditions that might interfere with sighting a line by a transit telescope.

A laser can frequently be set up on a bracket and aimed properly on (or through) a pair of targets and then be left in position unattended as an alignment device.

Some lasers are attachable to a transit so they shine parallel to the line of sight of the transit and the transit circles can be used in pointing the beam to line and grade. Others are clamped to the transit telescope in such a manner that it is possible to direct the laser beam through the telescope of the transit. This is done by using a double-mirror arrangement that swivels into position at the eye end of the transit telescope,

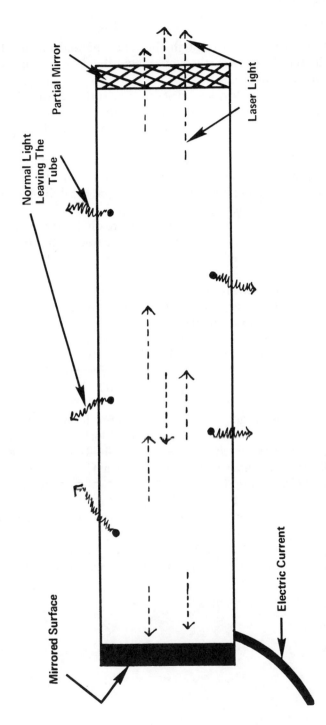

Fig. B.1 Schematic of Simplified Laser-generating Tube.

287

once the operator has sighted through the telescope and aligned the instrument. Still other lasers have their own mounting brackets and circles, and do not need a transit.

The laser is cited in the text as a distance-measuring device. The laser beam can be modulated and used as a light source in a type of electronic distance-measuring (EDM) equipment, which measures the time lapse between sending a signal and receiving it back from a prism reflector (retroreflector). Sections 5.15 and 5.16 are devoted to this use of the laser. Many manufacturers are producing this type of EDM equipment now, and it gives great promise of being a major breakthrough in distance measurement and construction layout.

Distances as short as 10 m are measurable virtually better than by tape by many of these systems, and the usual construction distances between 100 and 1500 ft are assuredly more accurately measurable than ever before.

Other laser uses not specifically mentioned elsewhere are aligning of stakes for airport paving, paint-striping of a runway, setting posts for a median barrier on a highway, or the laying of railway track. Many similar uses will suggest themselves, to be sure.

Power for lasers is minimal; typically a construction laser can operate all day on a 12-V automobile battery without recharging. Provision normally exists for such direct-current operation or for alternating current at 110 V.

There are also viewing goggles with interference filters that pass light at only the very narrow laser wavelength but block virtually all background light. They thus greatly enhance the visibility of the laser beam and are particularly valuable in locating the beam at long distances in very bright sunlight.

The laser has the virtue of being quick, accurate, inexpensive, versatile, visible, and easy to use. The construction industry can expect to see its use increase greatly.

It should be pointed out that some states (e.g., N.Y.) require certification of laser operators, even registration of laser equipment, as a safety precaution.

C

Three-Wire Leveling

A method of differential leveling that is speedy, accurate, and self-checking is that using the three cross wires of the level telescope to read three readings on the rod. It gives the equivalent of three level runs, it furnishes an estimate of distance to each rod, and it enables one to spot a mistake in a reading at once before the rodman or the instrumentman moves from position. The notekeeper has to keep up with his subtractions, and tends to be the controlling person in the field party because he can see verification or lack of it. He can and should guide the rodman, also, to adjust his distance from the instrument and to keep backsight distance about equal to foresight distance. A sample of notes taken by this method is given; there are some variations that can be worked out, especially if one is using a yard-rod or a metric rod, but what is shown here for a foot-rod is very acceptable practice. A suggested item of information is a note as to what is the point (by description, by number or letter designation, etc.) that is being sighted at the moment.

BACKSIGHTING			FORESIGHTING		
+ ROD	STAD.	MEAN.	- ROD	STAD.	MEAN.
2.593			7.564		
	169			260	
2.424		2.4253	7.304		7.3030
	165			263	
2.259			7.041		
	(334)	(2.4253)		(523)	(7.3030)
3.821			7.173		
	206			114	
3.615		3.6160	7.059		7.0587
	203			116	
3.412			6.943		
	(743)	(6.0413)		(753)	(14.3617)
8.519			7.306		
	188			181	
5.331		5.3320	7.125		7.1237
	185			185	
5.146			6.940		
	(1116)	(11.3733)		(1119)	(21.4854)
(34.120)			(64.455)		
÷3=			÷3=		
+11.373		+11.3733	-21.485		-21.4854
					+11.3733

DIFFERENCE OF ELEVATION = -10.1121

Fig. C.1 Simple 3-Wire Level Notes.

D

Topographic Maps, Contours, and Photogrammetry

Topographic Maps

A topographic map is a graphic representation of selected man-made and natural features of a part of the earth's surface plotted to a definite scale. Note that a map is drawn to scale; it thus differs from a sketch, which need not be to scale and thus cannot be used to scale from. When a map has elevation and contour information it is called a topographic map, since the vertical dimension portrayal gives the picture of the undulations and slopes and levels of the terrain—the topography. The distinguishing characteristic of a topographic map is this portrayal of the shape and elevation of the terrain, showing the location and shape of the mountains, valleys, and plains, the network of streams and rivers, and the principal works of man.

Topographic maps record in convenient, readable form the physical characteristics of the terrain as determined by engineering surveys and measurements. Measurements must be made to plot sufficient points on paper to serve the purpose for which the map is made. One principal purpose is to permit planning and designing of civil engineering and architectural works, and both proper scale and proper contouring are important.

Map Scale

Topographic maps depict the earth's surface to correct scale, and map scale is selected to make the finished product capable of serving its purpose. To map at too large a scale makes the topographic map too costly; to map at too small a scale makes the product essentially useless for its intended function. For highway planning, one may require a map at 1 in. = 200 ft (1:2400), whereas for detailed design of the highway, one may need a scale of 1 in. = 40 ft (1:480). For building sites, a scale 1 in. = 20 ft (1:240) is usual. On the other hand, preliminary planning over large areas may need a much smaller-scale map, say, at 1 in. = 2000 ft (1:24,000).

Contours

Topography, the shape of the earth's real surface, the ground, is generally shown by contours on maps. Contour lines are lines on the map connecting points of equal elevation. They are handy for computation, as also for conveying the proper understanding of the nature of the terrain. Contours can be drawn by discovering elevations on a grid-based ground pattern, or from the elevations of any scattered but thorough array on the ground of points whose position can be established on paper. It is possible, too, that contour lines can be plotted by use of steroscopic pairs of aerial photographs, using principles of photogrammetry—and most acreage today is mapped that way.

Contour lines can also be drawn on an existing plan if enough elevations are found and spotted on the plan. This can be a combined leveling and taping operation (even pacing) with the points being plotted in the field as the survey progresses. Leveling is extended from existing bench marks, perhaps through several turning points, and, for security, returning to known elevations: the circuit should be closed. Then, as discussed in Sections 2.20 and 2.23, significant elevations are noted directly on the existing plan so contours can be interpolated. A sketch of such an operation is shown in Fig. D.1, with the contour lines already sketched in. This is a typical instance of a need for additional topographic information to fill in some missing contours in an area. In this sketch the selected control points used in the contouring are shown as dots, though these would normally be erased once they are no longer needed.

Plotting elevations and drawing contours in the field in this manner finds application frequently when aerial photographs are used as a base. The photo, perhaps enlarged and printed faintly, serves as the base on which to draw, principally to fill in areas where the photogrammetric operator is unsure and desires checking, or at very critical places, or where tree cover has obliterated his view of the ground. The base photo

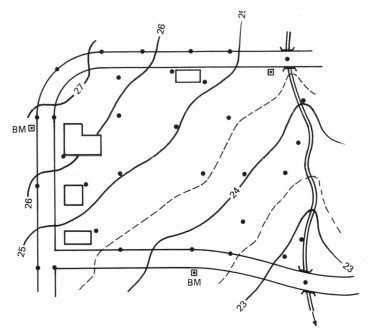

Fig. D.1. Adding contours by random control points on an existing plan.

contains whatever elevation control information is available, and the fieldmen then essentially fill in the gaps by finding their position on the photo and working from there.

Topographic Mapping by Photogrammetry

Mapping procedures have changed considerably since the time when all topographic maps were sketched by hand in the field, using an alidade and plane table. Prohibitive costs by the older methods of compilation often restricted their use to large projects and special purposes. Today, most maps are compiled in the office by photogrammetric methods, using stereoscopic plotting instruments with aerial photographs taken from 1000 to 60,000 ft above the ground. On the three-dimensional optical model of the terrain that is viewed by the operator of a stereoscopic plotting instrument, precise measurements can be made for accurate delineation of contours, drainage, woodland, and man-made culture. A minimal amount of field measurement is required to provide control for the mapping. The introduction in recent years of photogrammetric methods and equipment has made possible more speed and greater accuracy than ever before, while reducing the cost of maps to a point where every type of enterprise can avail itself of mapping services when required. Figs. D.2 through D.5 show varied uses of photogrammetric topographic maps. Fig. D.6 is of interest as a sample of a map with contours being plotted by an electronic computer.

The topographic map serves many needs; its value to mankind in general and to engineers in particular has long been recognized. It has many uses as a fundamental tool in planning, designing, and executing airports, highways, dams, pipelines, transmission lines, industrial plants, and countless other types of construction. While the layout man may need specific elevations, the planner works with topographic maps that show the area of the project with the sense of its shape. It is a question of which scale best serves each one's need, but the surveyman has had a hand in obtaining the information that led up to making any topographic map, having done the measurements for it. He has come full circle when the planner-designer gives him a layout task based on the map made from his previous measurements.

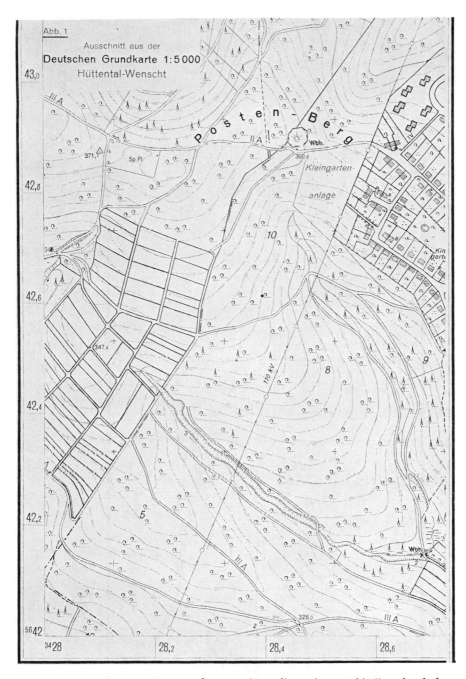

Abb. 1
Ausschnitt aus der
Deutschen Grundkarte 1:5 000
Hüttental-Wenscht

Fig. D.2 Typical topographic map for general use (here photographically reduced about 27% to 1:6900).

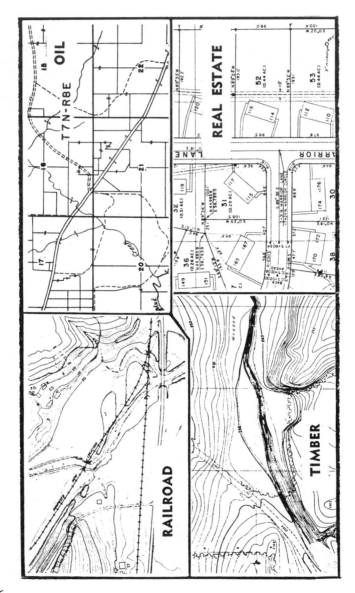

Fig. D.3 Uses of topographic maps.

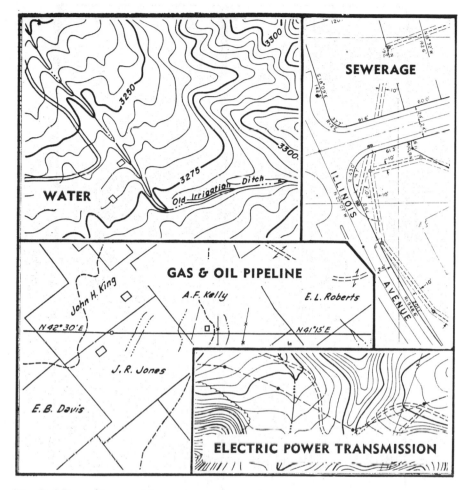

Fig. D.4 Uses of topographic maps.

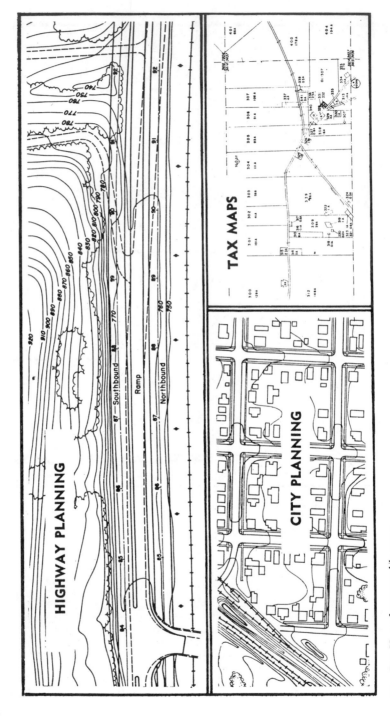

HIGHWAY PLANNING

TAX MAPS

CITY PLANNING

Fig. D.5 Uses of topographic maps.

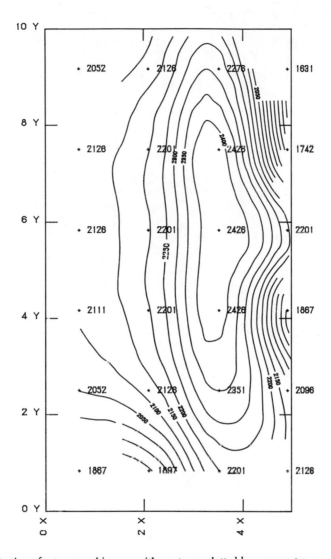

Fig. D.6 Portion of a topographic map with contours plotted by a computer.

Index